Copyright

ISBN 978-1-7389501-9-5

`academy@existsforall.com`

Acknowledgements

"At the age of eleven, I began Euclid, with my brother as tutor. This was one of the great events of my life, as dazzling as first love. I had not imagined there was anything so delicious in the world. From that moment until I was thirty-eight, mathematics was my chief interest and my chief source of happiness."

– *Bertrand Russell, Autobiography*

"Mathematicians, like Proust and everyone else, are at their best when writing about their first love."

– *Gian-Carlo Rota, Discrete Thoughts*

I express my gratitude to:

- The Almighty Creator, for providing me with this blessed and privileged life.
- My parents, for financing my mathematical education, and for supporting me during the time that this book was written.
- My students, for asking wonderful questions.
- My friends, for their companionship and for listening to me talk about mathematics.
- Euclid, for writing the *Elements*, which showed the world the meaning of eternal rigour.

Special thanks are extended to Warren Bei for reading the the first edition's manuscript and offering numerous suggestions, many of which were implemented. Any remaining mathematical errors or mistakes in the typesetting are my responsibility alone.

No artificial intelligence was used in the writing of the material for this book. Occasional assistance was sought from large language models for surmounting typesetting and formatting challenges in LaTeX.

Contents

Preface **3**

1 Addition and Subtraction **5**
1.1 Cardinality . . . 5
1.2 Lists and Multisets . . . 10
1.3 Basic Principles . . . 13

2 Pigeonhole Principle **19**
2.1 Injections and Surjections . . . 19
2.2 Extending Pigeonhole . . . 23

3 Multiplication and Division **26**
3.1 Multiplication Principle . . . 26
3.2 Correspondence Principle . . . 31

4 Double Counting **38**
4.1 Walking on Blocks . . . 38
4.2 Forming Committees . . . 42

5 Algebraic Counting **47**
5.1 Binomial and Multinomial Expansions . . . 47
5.2 Generating Functions . . . 49
5.3 Direct Computation . . . 55

6 Inclusion-Exclusion **60**
6.1 General Formulas . . . 60
6.2 Applications of PIE . . . 64

7 Distributions **68**
7.1 General Problem and Easy Cases . . . 68
7.2 Compositions . . . 70
7.3 Partitions . . . 75

8 Graph Theory **78**
8.1 Trees . . . 78
8.2 Planar Graphs . . . 83
8.3 Ramsey Theory . . . 88

9 Probability **98**
9.1 Probability Spaces . . . 98
9.2 Almost . . . 104
9.3 Conditional Probability . . . 106
9.4 Expected Value . . . 111

10 Classic Recursions **117**
10.1 Recursive Counting . . . 117
10.2 Fibonacci Numbers . . . 118
10.3 Catalan Numbers . . . 121

11 Linear Recurrences **125**
11.1 Non-homogeneous 125
11.2 Homogeneous 127

12 Group Theory **133**
12.1 Groups and Burnside 133
12.2 Counting Necklaces 136

Appendices **141**

A Matrices **142**

B Solutions **144**

List of Symbols **176**

Bibliography **178**

Index **179**

Preface

> "In studying a philosopher, the right attitude is neither reverence nor contempt, but first a kind of hypothetical sympathy, until it is possible to know what it feels like to believe in his theories, and only then a revival of the critical attitude, which should resemble, as far as possible, the state of mind of a person abandoning opinions which he has hitherto held. Contempt interferes with the first part of this process, and reverence with the second. Two things are to be remembered: that a man whose opinions and theories are worth studying may be presumed to have had some intelligence, but that no man is likely to have arrived at complete and final truth on any subject whatever."
>
> – *Bertrand Russell, A History of Western Philosophy*

Mathematics is the study of ultimate regularity. Regularity entails order or predictability. Its antithesis is chaos. When there is regularity, there are discernible objects at play. In other words, there is structure. Wherever there is structure, there is symmetry. Symmetry means that, while one aspect of the object changes, another remains unchanged. The present book is an effort to rigorously systematize and provide an exposition of those aspects of elementary counting that appear in school curricula and mathematical contests. In the course of writing, it became evident that there are three recurring themes among the proof techniques used, all of which are forms of symmetry:

A. The discrete Fubini's principle instructs us to write the same object in two different ways. There are many examples of how we have applied this principle:

   1. There is a whole chapter on double counting (Chapter 4), which is about counting one set in two ways in order to produce a combinatorial identity.
   2. It is used to find a formula for the sum of the pth powers of the first n positive integers (Example 5.3).
   3. We use it to find a formula for the sum of the elements of a row of Pascal's triangle where the lower entries of the binomial coefficients are all multiples of some given integer (Theorem 5.23).
   4. Two different ways of performing a binomial expansion allow us to prove Vandermonde's identity (Theorem 5.9).
   5. A variation allows us to develop a proof of a generalization of the principle of inclusion-exclusion (Corollary 6.4).
   6. It is used to solve homogeneous linear recurrences where the characteristic polynomial has all distinct roots (Theorem 11.10) or all equal roots (Theorem 11.13).
   7. It plays a key role in proving Burnside's lemma (Theorem 12.7).

B. Antisymmetry in a partial order is a powerful method of proof that lets us break down the strong notion of equality into the conjunction of two individually weaker statements. Examples of antisymmetry that we have used in this book are:

   1. A form of antisymmetry is given as the finite Schröder-Bernstein theorem (Theorem 2.8), which essentially says that if there exist injections $f : X \to Y$ and $g : Y \to X$ between finite sets, then there exists a bijection between the sets. The general case holds as well, that is for sets that are not necessarily finite.
   2. Set equality, as described above, is used to solve Bertrand's ballot problem (Theorem 9.7).
   3. Real number equality, as described above, is used to prove that the Calkin-Wilf tree forms a bijection with the positive rationals (Theorem 1.15). We also use it to prove the symmetry in Ramsey's theorem (Problem 8.38). The technique is often used in probability too, as exemplified several times in Section 9.2.

C. Applying an equivalence relation to a set groups the distinct elements into equivalence classes, where elements in a class are equated. This allows us to focus on the essential properties of objects which are preserved under the relation, and to choose a convenient canonical representative of each class.

   1. While not exactly an equivalence relation, the property of equipotence on sets has the relevant features (see Lemma 1.2). It is useful throughout the study of counting. In particular, it leads to our first and most important combinatorial technique, the bijection principle (Theorem 1.10), which is used to prove other principles and solve many problems.

It is our hope that the reader will keep these proof techniques in mind while reading the book, and that the impression of the importance of symmetry will grow as the reader encounters the methods time and again.

The intended audience consists of students of math contests, competitions, and olympiads who want to take a rigorous second look at the results that they might be accustomed to taking for granted, and teachers, coaches, and trainers who want to reinforce their own understanding of what they teach.

Suggestions, comments, and error submissions would be greatly appreciated. These could include ideas for strengthening or generalizing theorems and other additional material, and fixing mathematical or typographical mistakes. Messages may be sent to

`academy@existsforall.com`

Samer Seraj
Mississauga, Ontario, Canada
June 6, 2026

Chapter 1

Addition and Subtraction

> "What was designated by [combinatorics] was a part of set theory, namely the theory of finite sets, and the problems of estimating the numbers of members of such sets, constructed by following various procedures."
>
> – *Jean Dieudonné, Mathematics - The Music of Reason*
>
> "God created infinity, and man, unable to understand infinity, had to invent finite sets."
>
> – *Gian-Carlo Rota, Discrete Thoughts*

Counting, or combinatorics, is the field of mathematics that is dedicated to counting how many elements are in a finite set, given a description of the set; a part of it is also about proving that a set is finite before counting it. As a prelude to combinatorics, we need to make it clear what is meant by "finite" and what is meant by "counting" the number of elements of a finite set [6]. Many of the finite sets that are studied in combinatorics are defined in terms of lists and multisets, and being able to determine the cardinalities of those sets results in building blocks for determining the cardinalities of more complicated sets. To this end, counting techniques need to be developed. We will encounter the first method, that is the bijection principle, while studying cardinality. Our next two counting techniques will correspond to the arithmetic operations of addition and subtraction. The former splits the counting into individual cases that are added up, and the latter subtracts the number of "opposite" possibilities from the number of total possibilities in an overarching set. We will also see the initial cases of the principle of inclusion-exclusion. As a warning, our treatment of combinatorics will be detailed and rigorous.

1.1 Cardinality

Definition 1.1. A non-empty set X is said to be **equipotent** to a non-empty set Y if there exists a bijection $f : X \to Y$. This relationship is denoted by $X \approx Y$. The non-existence of such a bijection f is denoted by $X \not\approx Y$.

A bijection is also called a "one-to-one correspondence," especially in combinatorial contexts, but we will avoid this term because it can be confused with one-to-one functions, which is a term for injections.

Lemma 1.2. Set equipotence satisfies the following properties for all non-empty sets X, Y, Z:

1. $X \approx X$

2. If $X \approx Y$, then $Y \approx X$; as such, it is not necessary to say which of X, Y is equipotent to the other, and we can instead say that X and Y are equipotent (that is, to each other).

3. If $X \approx Y$ and $Y \approx Z$, then $X \approx Z$.

Proof. The first property is true because the identity function $\text{Id}_X : X \to X$ is a bijection. The second property is true because every bijective function $f : X \to Y$ has an inverse $f^{-1} : Y \to X$ that is also bijective (this is proven without the axiom of choice in Volume 1). The third property is true because the composition of two bijections is a bijection, which the reader should be able to verify by showing that the composition of two injections is an injection and the composition of two surjections is a surjection. □

As the reader might recognize, these three properties are analogues of reflexivity, symmetry and transitivity in an equivalence relation. Unfortunately, we cannot call equipotence an equivalence relation because an equivalence relation may only be defined on a set, whereas the overarching collection in Lemma 1.2 is "the set of all possible sets," which is not allowed to exist, due to Russell's paradox from Volume 1.

Definition 1.3. A set X is said to be **finite** if either $X = \varnothing$, or if X is non-empty and there exists a positive integer n such that $X \approx [n]$. Recall that $[n] = \{1, 2, \ldots, n\}$. If X is not finite, it is said to be **infinite**.

Of course, it is easy to define that the "number of elements" in the empty set is 0. We would also like to say that if X is a non-empty finite set such that $X \approx [n]$, then the "number of elements" in X is n. The problem is there could exist an integer $m \neq n$ such that $X \approx [m]$ as well, which would interfere with the definition. It turns out that such an integer m does not exist, but we will need to prove this. Painstakingly, we will do this now by working with sections $[n]$ of $\mathbb{Z}_+$. It will be well worth the effort, as the results that we will develop along the way will imply other seemingly obvious properties of finite sets. At the same time, readers who are in a hurry should feel free to skip to the definition of cardinality. As a secondary note, this process of treating the empty set separately from non-empty finite sets is not unique to defining the number of elements of a finite set, and will in fact permeate our study of finite sets.

Lemma 1.4. Let n be a positive integer, X be a non-empty set, and x_0 be an element of X. Then $X \approx [n+1]$ if and only if $X\backslash\{x_0\}$ is non-empty and $X\backslash\{x_0\} \approx [n]$.

Proof. For one direction of the proof, suppose $X\backslash\{x_0\}$ is non-empty and there exists a bijection $g : X\backslash\{x_0\} \to [n]$. Then we define $f : X \to [n+1]$ by

$$f(x) = \begin{cases} g(x) & \text{if } x \in X\backslash\{x_0\} \\ n+1 & \text{if } x = x_0 \end{cases},$$

which is easily seen to inherit the bijective property of g. Thus, $X\backslash\{x_0\} \approx [n]$ implies $X \approx [n+1]$.
For the other direction of the proof, suppose there exists a bijection $f : X \to [n+1]$. First we need to establish that $X\backslash\{x_0\}$ is non-empty, while being careful to not speak of the "number of elements" of any set since the notion is not yet well-defined. For reasons to be stated, there must exist some $k \in [n+1]$ such that $k \neq f(x_0)$. If it were otherwise, then all elements of $[n+1]$ would map to $f(x_0)$, and the injectivity of f would imply that x_0 is the only element of $[n+1]$. Then we would have $n+1 = 1$, contradicting the fact that $n+1 \geq 2$. Since f is surjective, there must be some element of X other than x_0 that f maps to k, which proves that $X\backslash\{x_0\} \neq \varnothing$.
Now we will prove the existence of a bijection $g : X\backslash\{x_0\} \to [n]$. If $f(x_0) = n+1$, then we can simply define $g = f|_{X\backslash\{x_0\}}$. For those unfamiliar with this notation, this is f restricted to the domain $X\backslash\{x_0\}$. If $f(x_0) \neq n+1$, the job is a bit harder but not much so. In this case, let $x_1 \neq x_0$ be the unique element of X that f maps to $n+1$. Then we can still define g to be a restriction to $X\backslash\{x_0\}$ but only after "swapping" the outputs of x_0 and x_1. That is, we define $g : X\backslash\{x_0\} \to [n]$ by $g = h|_{X\backslash\{x_0\}}$ where $h : X \to [n+1]$ is defined as

$$h(x) = \begin{cases} f(x) & \text{if } x \in X\backslash\{x_0, x_1\} \\ f(x_0) & \text{if } x = x_1 \\ f(x_1) = n+1 & \text{if } x = x_0 \end{cases}.$$

This function g is a bijection, the verification of which we leave to the reader. □

Lemma 1.4 allows us to switch between sets that differ by one element, which will turn out to be a useful tool in the inductive setting of the next lemma.

Lemma 1.5. Let n be a positive integer. Suppose X is a non-empty set such that $X \approx [n]$. If Y is a non-empty proper subset of X, then:

1. $Y \not\approx [n]$

2. There exists a positive integer $m < n$ such that $Y \approx [m]$.

Proof. We will proceed by induction on $n \geq 1$. In the base case $n = 1$, suppose X is a non-empty set such that $X \approx [n]$. Let x_0 be the only element of X so that $X = \{x_0\}$. In this case, the only proper subset of X is $\varnothing$, which not non-empty. Thus, the two assertions hold vacuously in the base case.
Now we assume that the two assertions are true for some positive integer n. Suppose X is a non-empty set such that $X \approx [n+1]$. Let Y be a non-empty proper subset of X (by Lemma 1.4, such a Y exists by removing any singleton from X, so it will not be a vacuous truth this time). An idea is to invoke the induction hypothesis after dropping an element $x_0 \in X$ such that $Y \backslash \{x_0\}$ is a proper subset of $X \backslash \{x_0\}$. This is satisfied by choosing x_0 to be any element of Y because elements x_1 of X that are outside Y remain inside $X \backslash \{x_0\}$ and outside $Y \backslash \{x_0\}$; x_0 exists because Y is non-empty and x_1 exists because Y is a proper subset of X. Before we can invoke the induction hypothesis, we use Lemma 1.4 to jump down by an element to get that $X \backslash \{x_0\}$ is non-empty and $X \backslash \{x_0\} \approx [n]$. Now either $Y \backslash \{x_0\} = \varnothing$ or $Y \backslash \{x_0\} \neq \varnothing$, and we will treat each case for each of the two assertions in the result.

1. If $Y \backslash \{x_0\}$ is empty, then by the contrapositive of one direction of Lemma 1.4, $Y \not\approx [n+1]$. If $Y \backslash \{x_0\}$ is non-empty, then the first part of the induction hypothesis says that $Y \backslash \{x_0\} \not\approx [n]$. Again, by the aforementioned contrapositive of one direction of Lemma 1.4, we now jump up by one element to get $Y \not\approx [n+1]$. This proves the first part of the assertion.

2. If $Y \backslash \{x_0\}$ is empty, then $Y = \{x_0\}$ and so $Y \approx [1]$. If $Y \backslash \{x_0\}$ is non-empty, then the second part of the induction hypothesis says that

$$X \backslash \{x_0\} \approx [n] \implies Y \backslash \{x_0\} \approx [p]$$

for some positive integer $p < n$. By one of the directions of Lemma 1.4, we can jump up by an element to get $Y \approx [p+1]$. So, we can define

$$m = \begin{cases} 1 & \text{if } Y \backslash \{x_0\} = \varnothing \\ p+1 & \text{if } Y \backslash \{x_0\} \neq \varnothing \end{cases}.$$

Either way, $m < n+1$ and $Y \approx [m]$, which completes the induction.

□

Lemma 1.6. If X is a non-empty finite set and Y is a non-empty proper subset of X, then $X \not\approx Y$.

Proof. Since X is a non-empty finite set, $X \approx [n]$ for some positive integer n. As Y is a non-empty proper subset of X, Lemma 1.5 says that $Y \not\approx [n]$. For contradiction, suppose $X \approx Y$. Then $Y \approx X$ and $X \approx [n]$ together imply that $Y \approx [n]$. This is a contradiction. □

Theorem 1.7. If X is a non-empty set and n, m are positive integers such that $X \approx [n]$ and $X \approx [m]$, then $n = m$.

Proof. Suppose X is a non-empty set and that $X \approx [n]$ and $X \approx [m]$ for some positive integers n and m. Then $[n] \approx [m]$. Suppose, for contradiction, that $m \neq n$. Then either $m < n$ and so $[m] \subsetneq [n]$, or $n < m$ and so $[n] \subsetneq [m]$. Either way, Lemma 1.6 implies that $[n] \not\approx [m]$, which contradicts $[n] \approx [m]$. □

Definition 1.8. If X is non-empty and finite, then we know by Theorem 1.7 that there exists a unique positive integer n such that $X \approx [n]$. The **cardinality** of X is n. If $X = \varnothing$ then we define its cardinality to be 0. The cardinality of X is denoted by $|X|$. The cardinality of a finite set X is informally known as "the number of elements in X" and **counting** the number of elements in X means to find the cardinality of X. A set with cardinality k is called a **k-set**, and a subset with cardinality j of a set X is called a **j-subset** of X.

The reader is asked to take note that, in many contexts, one is asked to find the number of elements of a set, but if one intends to be careful, it is necessary to first check that the set in question is in fact finite. We will observe this formality for some time and then more or less drop it when the point has been driven home that it is a step that is technically necessary but that it is usually hidden behind the argument that is used to actually do the counting.

Problem 1.9. Let X be a non-empty finite set and Y be a proper subset of X. Show that Y is finite and $|Y| < |X|$.

A cornerstone of combinatorics, called the bijection principle, asserts along the line that if there is a bijection between two finite sets, then the two sets have the same cardinality. This principle is not ordinarily encountered early in one's combinatorial education. Instead, it is presented later on when ingenious bijections, in conjunction with previously established counting techniques, sometimes allow for the determination of the cardinality of a finite set. The issue is that we have been unable to establish those basic counting techniques without using the bijection principle, albeit in relatively simple ways. Thus, we have decided to not postpone the introduction of the bijection principle and instead make it our very first counting principle.

Theorem 1.10 (Bijection principle)**.** Suppose X and Y are non-empty sets such that at least one of them is finite. Then $X \approx Y$ if and only if both X, Y are finite and $|X| = |Y|$.

Proof. This is a simple matter of unwrapping definitions. For both directions of the proof, we will assume that X and Y are non-empty and that, without loss of generality, X is finite; there is a symmetrical proof if we assume that Y is finite instead of X. For one direction, suppose $X \approx Y$. Since X is finite and non-empty, there exists a positive integer n such that $X \approx [n]$. Then $Y \approx [n]$ as well, proving that Y is finite and $|Y| = n = |X|$. Conversely, suppose X, Y are both finite and $|X| = |Y|$. Then there exists a positive integer m such that $|X| = m = |Y|$. Thus, $X \approx [m]$ and $Y \approx [m]$, implying $X \approx Y$. □

Problem 1.11. Prove that adding in or dropping an element from a finite set changes its cardinality by 1. More precisely:

1. Let X be a finite set and x_0 be an element that is not in X. Use the bijection principle to show that $X \cup \{x_0\}$ is finite and $|X \cup \{x_0\}| = |X| + 1$.

2. Let Y be a non-empty finite set and let y_0 be an element of Y. Use the first part of the problem to show that $Y \backslash \{y_0\}$ is finite and $|Y \backslash \{y_0\}| = |Y| - 1$.

Definition 1.12. A set X is said to be **countably infinite** or **countable** if X is non-empty and $X \approx \mathbb{Z}_+$.

Problem 1.13. Show that the integers are countable.

Theorem 1.14. The rationals are countable.

Proof. It suffices to prove that the positive rationals are countable because then we can start with 0, and toggle between positives and negatives, just like in the solution to Problem 1.13. Each positive rational may be written in the form $\frac{a}{b}$ for positive integers a and b. Thus, we can find a bijection from the positive integers $\mathbb{Z}_+$ to the positive rationals $\mathbb{Q}_+$ using the "Cantor snake" below, where the rationals are organized into an infinite matrix.

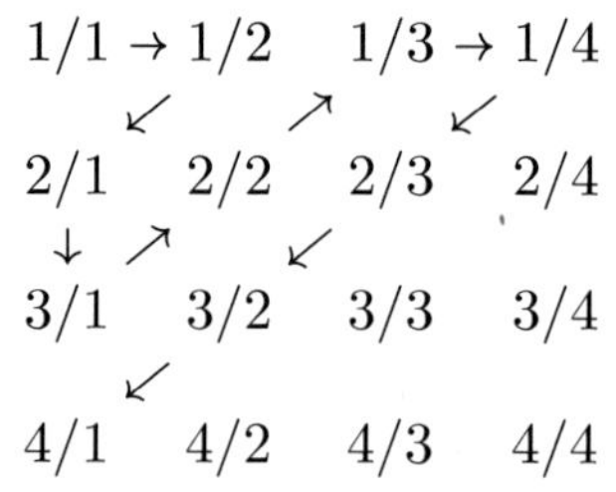

The only potential issue is that the fractional representation $\frac{a}{b}$ of positive rationals is not unique. This is not an actual problem because when we are going through the above infinite matrix using the Cantor snake, we may simply skip over any fraction that is equal to a fraction that has previously been included. □

Theorem 1.15 (Calkin-Wilf tree)**.** An infinite tree diagram is drawn so that:

- There is a distinguished node, called the root, which has no parent.
- Every node has two children: a left child and a right child.

- The root node has value 1.
- If a node has value v, then its left child has value $\dfrac{v}{v+1}$ and its right child has value $v+1$. As a result, it is clear inductively that all values in the tree are rational.

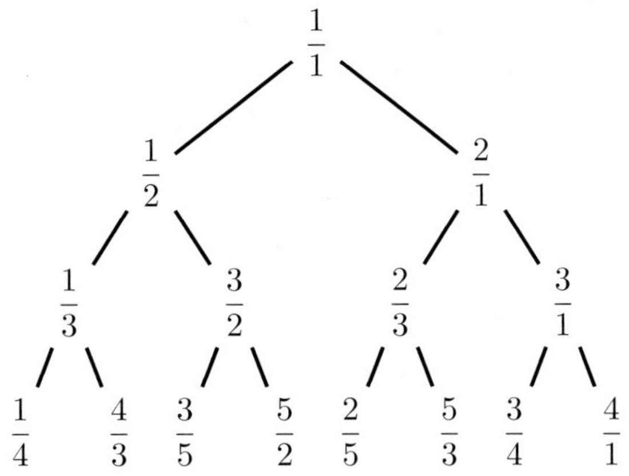

By iterating through one level at a time, where each level is finite, all positive rational numbers are uniquely enumerated. To be precise, every positive rational number appears exactly once in the tree, This avoids the redundancies in the Cantor snake.

Proof. We will proceed along the lines of the proof in Calkin and Wilf's original paper [2]. Suppose a node has value $\frac{r}{s}$ in lowest terms. Then its left child has value

$$\frac{\left(\frac{r}{s}\right)}{\frac{r}{s}+1} = \frac{r}{r+s}$$

and its right child has value

$$\frac{r}{s}+1 = \frac{r+s}{s}.$$

By the faux-Euclidean algorithm from Volume 3,

$$\gcd(r, r+s) = \gcd(r, s) = 1,$$
$$\gcd(r+s, s) = \gcd(r, s) = 1,$$

so $\dfrac{r}{r+s}$ and $\dfrac{r+s}{s}$ are in least form as well. Also, note that the left child always has a value strictly less than the value of its parent, and the right child always has a value strictly greater than the value of its parent; by transitivity, the left child has value strictly less than the right child.
Secondly, we note that if $w = \frac{r}{s}$ is the value of a left child and its parent has value v, then

$$w = \frac{v}{v+1} \implies v = \frac{w}{1-w} = \frac{r}{s-r}.$$

Similarly, if $w = \frac{r}{s}$ is the value of a right child and its parent has value v, then

$$w = v+1 \implies v = w-1 = \frac{r-s}{s}.$$

These inverse formulas will be helpful in proving that every positive rational appears at least once in the tree and at most once in the tree:

1. Suppose, for contradiction, that some positive rational in reduced form does not appear in the tree. By the well-ordering principle, we collect such rationals with the least denominator, and then select the one, say $\frac{r}{s}$, among them with the least numerator. As is common in number theory proofs involving the well-ordering principle, we will aim to construct an element that contradicts this minimality. Since the root, with value 1, definitely appears, we know $r \neq s$, so we are dealing with a non-root node, meaning it has a parent. By $r \neq s$, it must be true that $r < s$ or $r > s$.

- If $r < s$, then we are working with a left child. In this case, its supposed parent $\dfrac{r}{s-r}$ cannot exist either, because it would give birth to a left child with the forbidden value $\dfrac{r}{s}$. This contradicts the minimality of the denominator of $\dfrac{r}{s}$ among those reduced fractions that do not appear.
- If $r > s$, then we are working with a right child. In that case, its supposed parent $\dfrac{r-s}{s}$ cannot exist because it would have a right child with the impossible value $\dfrac{r}{s}$. This contradicts the minimality of the numerator of $\dfrac{r}{s}$ among those reduced fractions that do not appear and have denominator s.

Thus, with a contradiction either way, every reduced fraction must appear at least once in the tree.

2. Suppose, for contradiction, that some positive rational in reduced form appears at least twice in the tree. As in the previous step, we use the well-ordering principle to collect such rationals in the tree with the least denominator, and select the one, say $\dfrac{r}{s}$, among them with the least numerator. Since left and right siblings (that is, children descended from the same parent) cannot have equal values, the two occurrences of $\dfrac{r}{s}$ must have different parents. So $\dfrac{r}{s}$ is not 1, leading to $r \neq s$. Then $r < s$ or $r > s$.

   - If $r < s$, then both occurrences are at left children, each of which has a distinct parent with value $\dfrac{r}{s-r}$. This contradicts the minimality of the denominator of $\dfrac{r}{s}$ among those reduced fractions that appear more than once.
   - If $r > s$, then both occurrences are at right children, each of which has a distinct parent with value $\dfrac{r-s}{s}$. This contradicts the minimality of the numerator of $\dfrac{r}{s}$ among those reduced fractions that appear more than once and have denominator s.

   Either way, we have a contradiction. Thus, every reduced fraction appears at most once in the tree.

By antisymmetry, every positive rational appears exactly once in the tree. Therefore, we can enumerate the rationals without any repeated entries by completing one row at a time, from the top row (the root) downwards. The author was alerted about the existence of this result by Warren Bei. □

Problem 1.16. Let X be a non-empty set. Show that if X is countably infinite, then X is infinite, as in not finite. (Thank goodness, otherwise the name "countably infinite" would be confusing!)

In Volume 3, we will show that the real numbers are *uncountable* in the sense that, despite being $\mathbb{R}$ infinite, there does *not* exist a bijection $f : \mathbb{Z}_+ \to \mathbb{R}$.

1.2 Lists and Multisets

We briefly touched on lists when looking at indexing in Volume 1. Informally, a list is a collection, where the elements are ordered according to a section $[n]$ of $\mathbb{Z}_+$ and repeated elements are allowed. In essence, a list is a collection-like structure that breaks both rules required to be a set. One might think of a list as a generalization of an ordered pair. The following is a formal definition.

Definition 1.17. A non-empty **list** is a function $f : [n] \to S$ for some positive integer n and some non-empty set S. The **indices** of f are the elements of $[n]$, and the **entry** at index i is $f(i)$ (we will often use the word "entry" instead of "element" for lists). The **length** of f is n, which is also said to be the "number of entries" in f. A non-empty list of length n is called an **n-tuple**. We write out the entries of f by enclosing them within parentheses in order from left to right like $(f(1), f(2), \ldots, f(n))$. The **empty list** is defined to be the list with no elements in it and it is denoted by $()$. Two lists are said to be **equal** if they are equal as functions, meaning they have the same domain (and so have the same length) and each function maps the same element of the domain to the same output.

Example. The list (a, b, c, d) has length 4. Its first entry is a, second entry is b, third entry is c, and fourth entry is d.

Let us see some examples of writing out all elements of a described set. As a result, we will gain an appreciation of why it is desirable to develop more sophisticated counting techniques that bypass manual counting. However, we note that it is sometimes necessary to know all possibilities, so being able to exhaustively write out the elements of a set is not a superfluous skill.

Example 1.18. A **palindrome** is a list of characters, such as numbers and letters, such that it reads the same forwards as backwards. For example, $a5a$ and $12c21$ are alphanumeric palindromes, where we have dropped the parentheses and commas of list notation without introducing ambiguity. Write out all two-character palindromes consisting of digits from the set $[9]^* = \{0, 1, 2, 3, 4, 5, 6, 7, 8, 9\}$.

Solution. We can simplify the problem by noticing that both characters in a two-character palindrome are the same. So we just need to know the possible single characters, and then the two-character palindromes will each be a character followed by its duplicate. The possible characters are the elements of the set given to us, so the set of two-character palindromes is

$$\{00, 11, 22, 33, 44, 55, 66, 77, 88, 99\}.$$

Formally, we have used a bijection between the given set of single digits and the described set of two-character palindromes. As a side note, if we had been asked for palindromes that are two-digit **integers**, then 00 would not have been a valid member of the set. □

Problem 1.19. Write out all 3-character lists abc where a, b, c are chosen from the set

$$[9]^* = \{0, 1, 2, 3, 4, 5, 6, 7, 8, 9\}$$

and $b = a + 1$ and $c = b + 1$.

Problem 1.20. Write out all increasing lists of 3 different numbers chosen from the set

$$[5] = \{1, 2, 3, 4, 5\}.$$

By increasing, we mean that the list abc satisfies $a < b < c$.

Informally, a **multiset** is a collection, where elements are not ordered, but repeated elements are allowed. In a sense, a multiset is a structure that lies somewhere between a set and a list. The following is a formal definition.

Definition 1.21. A non-empty **multiset** is a function $f : S \to \mathbb{Z}_+$, where each element of a non-empty set S, called the **support** of the multiset, maps to a positive integer called the **multiplicity** of the element. The multiplicity of an element is interpreted as the number of times the element appears in the multiset. Though it is not standard, we write out the elements of a multiset by enclosing them within angular brackets in any order from left to right like $\langle \ldots \rangle$. The **empty multiset** is defined to be the multiset with no elements in it, and it is denoted by $\langle\rangle$ (it can have any non-empty support, but the multiplicity of each element of the support is 0). A multiset is said to be **finite** if it is either empty or it is a non-empty multiset whose support S is finite (note that this means that the sum of all multiplicities is then finite too). The **cardinality** of the empty multiset is 0, the cardinality of a non-empty finite multiset is the sum of the multiplicities of the elements of its support, and we do not define the cardinality of a non-finite multiset; in particular the cardinality or "number of elements" of a finite multiset does *not* refer to the cardinality of the support. A multiset of cardinality n is said to be an n**-multiset**. Two multisets are said to be **equal** if each element that appears in a multiset appears the same number of times in the other multiset.

Example. Multisets are underused in mathematics, but there are certain collections that are naturally expressed as multisets. For example, a consequence of the fundamental theorem of algebra from Volume 1 is that the complex roots of a degree n polynomial with complex coefficients form a multiset of cardinality n. Similarly, those familiar with number theory might recognize that the fundamental theorem of arithmetic from Volume 3 says that any integer greater than 1 is equal to the product of the elements of a unique finite multiset of primes. These two results are rarely stated using multisets.

Problem 1.22. Write out all 4-multisets such that each element is 0 or 1.

As this is a discussion about multisets, we will take this as an opportunity to provide the reader with an overview of essential knowledge about elementary statistical measures of non-empty finite multisets of real numbers.

Definition 1.23. The **average** of a non-empty finite multiset whose support is a set of real numbers is the sum of all the elements of the multiset divided by the cardinality of the multiset. That is, the average of the multiset $\langle a_1, a_2, \ldots, a_n \rangle$ is

$$A = \frac{a_1 + a_2 + \cdots + a_n}{n}.$$

There are two other forms of this equation that are illuminating in their own right.

- The first alternative form is

$$a_1 + a_2 + \cdots + a_n = n \cdot A,$$

which emphasizes that the sum of n copies of the average gives the sum of the elements.

- The second alternative form is

$$(a_1 - A) + (a_2 - A) + \cdots + (a_n - A) = 0,$$

which emphasizes that the fact that the elements balance around the average. Assuming the elements are not all equal to the average, some are above it and some are below it, and the sum of the differences $a_i - A$ cancel out to a total of 0.

Theorem 1.24. If A is the average of a non-empty finite multiset M, whose support is a set of real numbers, then

$$\min(M) \leq A \leq \max(M).$$

If A consists of only integers, we may strengthen the inequalities to say that

$$\min(M) \leq \lfloor A \rfloor \leq A \leq \lceil A \rceil \leq \max(M).$$

This simple idea is a surprisingly applicable tool in combinatorics.

Proof. Let the multiset be $M = \langle a_1, a_2, \ldots, a_n \rangle$. Let the minimum element be a_i and the maximum element be a_j. By the definition of minimum and maximum, it holds that

$$n \cdot a_i \leq a_1 + a_2 + \cdots + a_n \leq n \cdot a_j,$$

and dividing through by n yields the desired inequalities. The second set of inequalities follows from the first set, using the fact that a_i and a_j are integers. □

Definition 1.25. Given a non-empty finite multiset whose support is a set of real numbers, we can place the elements of the multiset in a list $(a_1, a_2 \ldots, a_n)$ where the a_i are in non-decreasing order, meaning $a_1 \leq a_2 \leq \cdots \leq a_n$. If the cardinality of the multiset is odd, then the **median** of the multiset is the number in the middle of the list. If the cardinality of the multiset is even, then the median of the multiset is the average of the two numbers in the middle of the list.

Definition 1.26. If the support of a non-empty finite multiset is a set of real numbers (this definition immediately generalizes to the support being any non-empty set), then the **mode** of the multiset is the preimage of the maximal element among those elements of $\mathbb{Z}_+$ whose preimage is non-empty. In simpler terms, these are the elements of the support that are tied for having the highest multiplicity in the multiset. If the mode is a singleton, we normally take the single element out of the set defined to be the mode and instead call this element the mode.

Definition 1.27. The **range** of a non-empty finite multiset of real numbers is the largest element of the support with non-zero image minus the smallest element of the support with non-zero image.

These are the most common basic ways of measuring a non-empty finite multiset whose support consists of real numbers. Each method is suited to different situations in the realm of elementary statistics.

1.3 Basic Principles

Sometimes, it is necessary to write out all elements of a finite set. If we are familiar with counting techniques, then we might be able to calculate the total number of items that should appear in our list of the elements of the set. As long as every element that we have written out fits the description of the set and is distinct, then we just need to hit the predicted total number of elements in order to be sure that the job is complete. However, if we do not know how many items to expect, then we need a different way of ensuring that we have not missed anything. One such method is to break up the possibilities into non-overlapping cases and that together cover all the possibilities, and where each case is small enough that we can be sure that we have not missed any possibility within the case.

Example 1.28. Write out all possible ways of ordering the characters a, b, c from left to right as a list, where each character is used exactly once.

Solution. We break up the possibilities into three cases: the leftmost character is a, b, or c.

1. If the leftmost character is a then the remaining characters are b, c or c, b, in those orders.

2. If the leftmost character is b then the remaining characters are a, c or c, a, in those orders.

3. If the leftmost character is c then the remaining characters are a, b or b, a, in those orders.

Thus, the set of possibilities is $\{abc, acb, bac, bca, cab, cba\}$, where we have avoided list notation, like (a, b, c), without ambiguity. □

Definition 1.29. A non-empty multiset of sets is said to be **pairwise disjoint** if its support is a set of sets that have pairwise empty intersection and the multiplicity of each non-empty element of the support is 1 (if $\varnothing$ is an element of the support then it has the freedom to have any positive integer as its multiplicity). We will not classify the empty multiset as pairwise disjoint so that there will never be a need to clarify that a pairwise disjoint multiset is non-empty; we can easily define otherwise, if desired. This entire definition will be more motivated in light of Definition 1.30.

Example. All multisets whose only element is just one set are pairwise disjoint. As a technical novelty, this means that, although we have defined that the empty multiset $\langle\rangle$ is not pairwise disjoint, the multiset containing the empty set $\langle\varnothing\rangle$ is pairwise disjoint.

Definition 1.30. A **generalized partition** of a set A is a pairwise disjoint multiset of subsets of A such that the union of these subsets is A. An ordinary **partition** of a set A is instead defined to be a *set* of *non-empty* pairwise disjoint subsets of A whose union is A. We have included the possibility of there being some copies of the empty set in a generalized partition because it is not always clear in combinatorics that a set is non-empty. A generalized partition or partition is said to be **finite** if has finitely many elements, including all multiplicities in the case of generalized partitions.

Example. A generalized partition of $[5] = \{1, 2, 3, 4, 5\}$ is

$$\langle\{1, 2\}, \varnothing, \{3, 4, 5\}, \varnothing\rangle$$

whereas a partition of the same set is $\{\{1, 2\}, \{3, 4, 5\}\}$. Note that $\{\{1, 2\}, \{3, 4, 5\}, \varnothing\}$ is not a partition of $[5]$ because the empty set cannot be an element of a partition. The multiset $\langle\varnothing, \varnothing\rangle$ is a generalized partition of $\varnothing$. However, there is no ordinary partition of $\varnothing$ because $\varnothing$ has no non-empty subset.

The idea of casework has an important analogue among the available techniques for proving that a set is finite and determining its cardinality. This is our next counting method, and it is exceedingly important because it will lead to several other counting principles.

Theorem 1.31 (Addition principle)**.** Let n be a positive integer. Suppose A is a set, and that we have an n-element generalized partition $\langle A_1, A_2, \ldots, A_n\rangle$ of A. Then all of the A_i are finite if and only if $A = A_1 \cup A_2 \cup \cdots \cup A_n$ is finite. If either case is assumed, then both A and all of the A_i are finite, and

$$|A| = |A_1| + |A_2| + \cdots + |A_n|.$$

This immediately implies that the result remains true if we instead work with an n-element ordinary partition

$$\{A_1, A_2, \dots, A_n\}$$

of a non-empty set A.

Proof. For one direction, suppose A is finite. Since each A_i is a subset of A, each A_i is itself a finite set. Now we will prove by induction on $n \geq 1$ that if all of the A_i are finite, then $A = A_1 \cup A_2 \cup \cdots \cup A_n$ is finite with cardinality

$$|A| = |A_1| + |A_2| + \cdots + |A_n|.$$

This will complete the proof because, if the initial assumption were instead that A is finite, then our first argument would show that all of the A_i are finite, at which point our induction proof could be invoked; of course, the part of the induction that proves that A is finite would be redundant in this case.

In the base case $n = 1$, the assertion is trivially true. With the base case established, suppose the proposition is true for some $n \geq 1$. Let $\langle A_1, A_2, \dots, A_n, A_{n+1} \rangle$ be an $(n+1)$-element generalized partition of a set A such that all of the A_i are finite sets. Then $\langle A_1, A_2, \dots, A_n \rangle$ is a pairwise disjoint multiset of n finite sets, making it an n-element generalized partition of $B = A_1 \cup A_2 \cup \cdots \cup A_n$. By the induction hypothesis, B is finite with cardinality

$$|B| = |A_1| + |A_2| + \cdots + |A_n|.$$

First we will show that B and A_{n+1} are disjoint. Indeed, the distributive property of set intersection over set union says that

$$\begin{aligned} B \cap A_{n+1} &= (A_1 \cup A_2 \cup \cdots \cup A_n) \cap A_{n+1} \\ &= (A_1 \cap A_{n+1}) \cup (A_2 \cap A_{n+1}) \cup \cdots \cup (A_n \cap A_{n+1}) \\ &= \underbrace{\varnothing \cup \varnothing \cup \cdots \cup \varnothing}_{n \text{ empty sets}} = \varnothing. \end{aligned}$$

Now let the finite disjoint sets B and A_{n+1} be represented as

$$\begin{aligned} B &= \{b_1, b_2, \dots, b_p\}, \\ A_{n+1} &= \{a_1, a_2, \dots, a_q\}. \end{aligned}$$

Then

$$B \cup A_{n+1} = \{b_1, b_2, \dots, b_p, a_1, a_2, \dots, a_q\},$$

which includes all the b_i and all the a_j because no pair of these elements are the same. We can relabel each a_j as b_{p+j} so that

$$B \cup A_{n+1} = \{b_1, b_2, \dots, b_p, b_{p+1}, b_{p+2}, \dots, b_{p+q}\},$$

which is in bijection with $[p+q]$. Thus, the disjoint union $B \cup A_{n+1}$ is finite and has cardinality

$$p + q = |B| + |A_{n+1}|.$$

Combining this result with the induction hypothesis, we get that $A = B \cup A_{n+1}$ is finite with cardinality

$$|A| = |B \cup A_{n+1}| = |B| + |A_{n+1}| = |A_1| + |A_2| + \cdots + |A_n| + |A_{n+1}|.$$

This completes the induction. □

Usage of the addition principle is often called "casework," with the sets in the generalized partition being called the "cases." Note that, in the addition principle, we could start off with a finite pairwise disjoint multiset of finite component sets and take their union to build an overarching set that turns out to be finite as well, or we can start off with an overarching finite set and split it into a generalized partition whose component sets are necessarily finite in number and individually finite. The direction in which the addition principle is applied depends on the situation. As I like to tell my students, mathematical equalities are "double-sided" so one should grow accustomed to transforming both left into right and right into left.

Example 1.32. There are four distinct green balls labelled $1, 2, 3, 4$ and three distinct orange balls labelled a, b, c. In how many ways can one green ball, followed by one orange ball, be selected?

Solution. We break the possibilities up into four cases, according to which of the four green balls is selected. In each of those cases, any one of the three orange balls can be selected. Thus, there are

$$3 + 3 + 3 + 3 = 12$$

ways to select one green ball and one orange ball. □

Unfortunately, applications of the addition principle are more interesting when there are other tools available. What often happens is that we break matters up into cases and then use more sophisticated tools on each case. Instead of showing more simple examples, we will work on developing more tools, the proofs of several of which will rely on the addition principle. For example, in combinatorics, it is sometimes easier to count the "opposite" of what we are asked to count, and subtract this opposite count from the "total" number of possibilities. Even though this involves two computations, one for the opposite and one for the total, it might still be easier than directly counting the number of elements in the given set.

Definition 1.33. Recall from Volume 1 that, given a set A and a set $\mathcal{U}$ such that A is a subset of $\mathcal{U}$, the **complement** of A with respect to $\mathcal{U}$ is the set

$$\overline{A} = \{x \in \mathcal{U} : x \notin A\}.$$

In other words, the complement of A with respect to $\mathcal{U}$ is every element of $\mathcal{U}$ that does not lie in A, which is the same as the set difference $\mathcal{U} \backslash A$. Within a specific context, it is often the case that $\mathcal{U}$ is implicitly understood and not specified, which is why it does not appear in the notation $\overline{A}$ and we can refer to "the" complement of A. The notation A^c is another way to denote the complement of A.

Theorem 1.34 (Subtraction principle)**.** Let A and $\mathcal{U}$ be sets such that $\mathcal{U}$ is finite and $A \subseteq \mathcal{U}$. Then A and $\overline{A}$ are both finite, and

$$|A| = |\mathcal{U}| - |\overline{A}|.$$

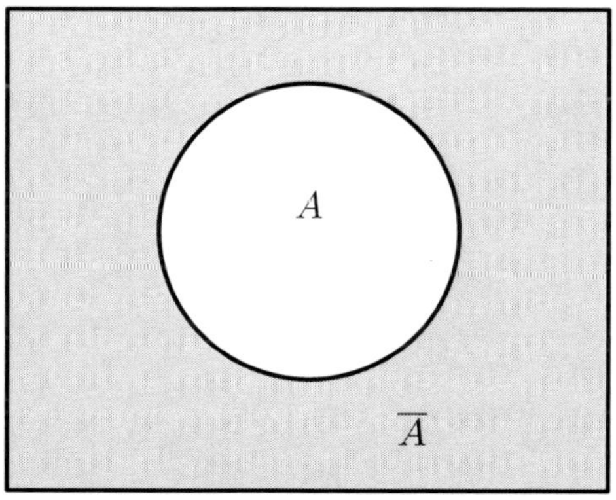

Proof. First, we rearrange the equation into the equivalent form

$$|A| + |\overline{A}| = |\mathcal{U}|.$$

This is reminiscent of the addition principle. Since $A \cap \overline{A} = \varnothing$ and $A \cup \overline{A} = \mathcal{U}$ (these two statements can be proven using logic but we will not do so, in order to avoid clutter), and A and $\overline{A}$ are both subsets of $\mathcal{U}$, we know that $\langle A, \overline{A} \rangle$ is a generalized partition of $\mathcal{U}$. Since $\mathcal{U}$ is finite, the addition principle says that A and $\overline{A}$ are both finite and $|A| + |\overline{A}| = |\mathcal{U}|$. □

Usage of the subtraction principle from combinatorics is called "complementary counting". As with casework, we will have to wait to see non-trivial examples of complementary counting. For now, the reader should take our word that this is indeed a valuable tool. The following is a variation that we call "symmetric complementary counting."

Problem 1.35. Suppose X is a non-empty finite set and that $\{A, B, C\}$ is a partition of X such that $B \approx C$. Show that A, B, C are all finite and

$$|B| = \frac{|X| - |A|}{2}.$$

This is a kind of pseudo-complementary counting that one might be able to apply when there is some form of symmetry between two elements of a 3-element partition of a finite set. For example, use this idea to find the probability that, when a blue standard die and a green standard die are rolled, the blue die rolls a strictly higher number than the green die.

In the addition principle, the component sets are required to be pairwise disjoint. In a general union of finitely many sets, it is not necessarily the case that the sets are pairwise disjoint. So, given a multiset $\langle A_1, A_2, \ldots, A_n\rangle$ of $n \geq 1$ finite sets, it is desirable to prove that $A_1 \cup A_2 \cup \cdots \cup A_n$ is finite and to find a formula for the cardinality of this union. This will rarely be the sum of the cardinalities of the component sets A_i because we have to account for the overlaps among them; the reader will be asked to put this statement into an inequality in Problem 1.39. Then comes a precise equation. The $n = 1$ case is trivial. We will determine the formula for $n = 2$ by an ad hoc method (Theorem 1.37) and $n = 3$ by using the $n = 2$ case (Theorem 1.40), both of which can be visualized using Venn diagrams. The general formula, called the principle of inclusion-exclusion, will be studied in Chapter 6.

Lemma 1.36. Let A and B be sets, not necessarily finite. Then the multisets

$$\begin{aligned}&\langle A\backslash B, B\rangle,\\ &\langle B\backslash A, A\rangle,\\ &\langle A\backslash B, A \cap B, B\backslash A\rangle\end{aligned}$$

are each a generalized partition of $A \cup B$.

Proof. We need to show that each of the three multisets is pairwise disjoint and that the union of each is $A \cup B$. As a fair warning, the formal proof will involve logical contortions. For the reader who is satisfied by Venn diagrams, there is no need to read it. For the reader who wants to see the logical argument, we have provided the details. As described in Volume 1, $\wedge$ stands for the logical conjunction "and," and $\vee$ stands for the logical disjunction "or." We will be using the logical distributive laws Volume 1.
In the first multiset, $A\backslash B$ contains no elements of B, so $(A\backslash B) \cap B = \varnothing$. For the union property,

$$\begin{aligned}(A\backslash B) \cup B &= \{x : x \in A \wedge x \notin B\} \cup \{x : x \in B\}\\ &= \{x : (x \in A \wedge x \notin B) \vee x \in B\}\\ &= \{x : (x \in A \vee x \in B) \wedge (x \notin B \vee x \in B)\}\\ &= \{x : x \in A \vee x \in B\}\\ &= A \cup B.\end{aligned}$$

A symmetric argument shows that in the second multiset,

$$\begin{aligned}&(B\backslash A) \cap A = \varnothing,\\ &(B\backslash A) \cup A = A \cup B.\end{aligned}$$

Now we tackle the third multiset. Since $A \cap B$ is a subset of B and we already know that $A\backslash B$ has an empty intersection with B, we find that $(A\backslash B) \cap (A \cap B) = \varnothing$. A symmetric argument gives $(B\backslash A) \cap (A \cap B) = \varnothing$. The final intersection $(A\backslash B) \cap (B\backslash A)$ is also empty because $A\backslash B$ contains no element of B, whereas $B\backslash A$ is a subset of B. For the union property,

$$\begin{aligned}(A\backslash B) \cup (A \cap B) \cup (B\backslash A) &= \{x : x \in A \wedge x \notin B\} \cup \{x : x \in A \wedge x \in B\} cup \{x : x \in B \wedge x \notin A\}\\ &= \{x : (x \in A \wedge x \notin B) \vee (x \in A \wedge x \in B) \vee (x \in B \wedge x \notin A)\}\\ &= \{x : (x \in A \wedge (x \notin B \vee x \in B)) \vee (x \in B \wedge x \notin A)\}\\ &= \{x : x \in A \vee (x \in B \wedge x \notin A)\}\\ &= \{x : (x \in A \vee x \in B) \wedge (x \in A \vee x \notin A)\}\\ &= \{x : x \in A \vee x \in B\} = A \cap B.\end{aligned}$$

Thus, the three multisets are all generalized partitions of $A \cup B$. □

Theorem 1.37 (Principle of inclusion-exclusion for two sets). If A and B are finite sets, then $A \cup B$ and $A \cap B$ are finite, and

$$|A \cup B| = |A| + |B| - |A \cap B|.$$

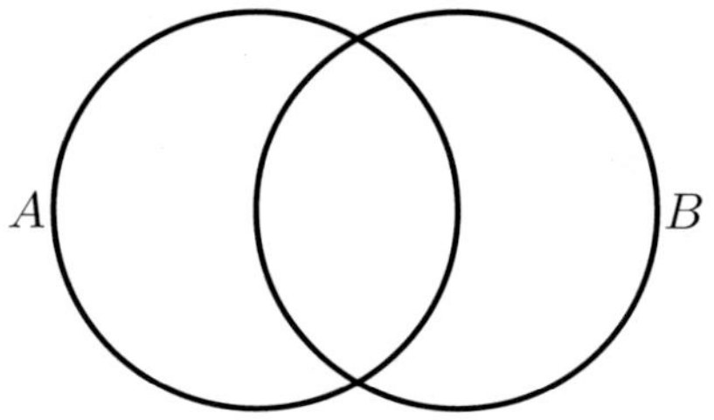

Proof. The main idea is to apply the addition principle to the three sets in Lemma 1.36, in order to obtain three equations that we can then add up. The formal details are given below.
Let A and B be finite sets. Then $A \backslash B$ and $A \cap B$ are subsets of A, and $B \backslash A$ is a subset of B. As subsets of finite sets, all three are finite sets. By Lemma 1.36, the multiset $\langle A \backslash B, A \cap B, B \backslash A \rangle$ is a generalized partition of $A \cup B$. Since the three component sets are finite, the addition principle says that $A \cup B$ is finite and

$$|A \backslash B| + |A \cap B| + |B \backslash A| = |A \cup B|$$

Similarly, since Lemma 1.36 says that $\langle A \backslash B, B \rangle$ and $\langle B \backslash A, A \rangle$ are generalized partitions of $A \cup B$, and all component sets and the overarching set $A \cup B$ are now known to be finite, the addition principle gives the equations

$$|A \cup B| = |A \backslash B| + |B|,$$
$$|A \cup B| = |B \backslash A| + |A|.$$

Adding up all three equations and cancelling terms that are common to both sides yields

$$|A \cup B| + |A \cap B| = |A| + |B|,$$

which is equivalent to the desired equation. □

Corollary 1.38. Let n be a positive integer. If $\langle A_1, A_2, \dots, A_n \rangle$ is a multiset of n finite sets, then

$$A_1 \cup A_2 \cup \cdots \cup A_n$$

is a finite set; this is more general than the analogous result in the addition principle because we are no longer assuming that the multiset of sets is pairwise disjoint. Moreover, any intersection of the type

$$A_{i_1} \cap A_{i_2} \cap \cdots \cap A_{i_k},$$

where $1 \leq k \leq n$ and $1 \leq i_1 < i_2 < \cdots < i_k \leq n$, is also finite.

Proof. An intersection of the given form is immediately seen to be finite because it is a subset of any of the component sets A_{i_j}, which are themselves finite. In fact, we just need one of the A_{i_j} to be finite.
The proof of the union being finite is by induction. The assertion is trivially true for $n = 1$. Then we assume that it holds for some positive integer n and let

$$\langle A_1, A_2, \dots, A_n, A_{n+1} \rangle$$

be a multiset of $n + 1$ finite sets. By Theorem 1.37, if A and B are finite sets, then $A \cup B$ is finite. Since A_{n+1} is finite and the induction hypothesis states that $A_1 \cup A_2 \cup \cdots \cup A_n$ is finite, it is true that

$$(A_1 \cup A_2 \cup \cdots \cup A_n) \cup A_{n+1}$$

is finite. This completes the induction. □

With Corollary 1.38 in place, it will be possible to speak of the cardinalities of such unions and intersections without first proving that they are finite sets. In some cases, we will not even preface a statement about cardinality by saying that the set in question is finite, if it is obvious enough that the set is finite.

Problem 1.39 (Union bound). Let n be a positive integer. Show that, if $\langle A_1, A_2, \ldots, A_n\rangle$ is a multiset of n finite sets, then

$$|A_1 \cup A_2 \cup \cdots \cup A_n| \leq |A_1| + |A_2| + \cdots + |A_n|,$$

with equality holding if and only if $\langle A_1, A_2, \ldots, A_n\rangle$ is pairwise disjoint.

Theorem 1.40 (Principle of inclusion-exclusion for three sets). If A, B, C are finite sets, then

$$|A \cup B \cup B| = |A| + |B| + |C| - |A \cap B| - |B \cap C| - |C \cap A| + |A \cap B \cap C|.$$

Note that each set whose cardinality is taken is finite, according to Corollary 1.38.

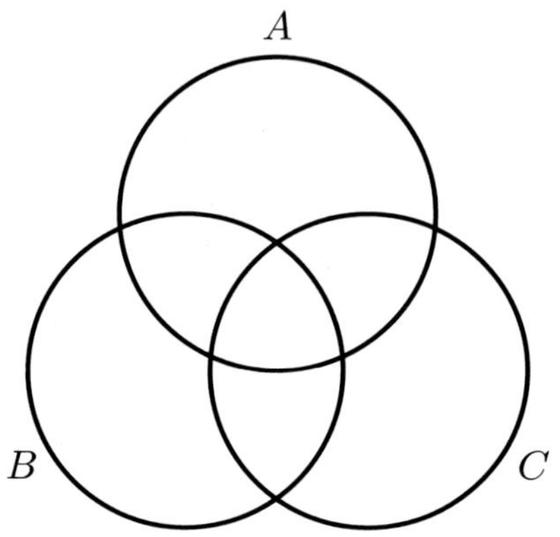

Proof. We will use the principle of inclusion-exclusion for two sets to prove this analogous result for three sets. By applying the formula for two sets several times, we get

$$\begin{aligned} |A \cup (B \cup C)| &= |A| + |B \cup C| - |A \cap (B \cup C)| \\ &= |A| + (|B| + |C| - |B \cap C|) - |(A \cap B) \cup (A \cap C)| \\ &= |A| + |B| + |C| - |B \cap C| - (|A \cap B| + |A \cap C| - |(A \cap B) \cap (A \cap C)|) \\ &= |A| + |B| + |C| - |B \cap C| - |A \cap B| - |A \cap C| + |A \cap B \cap C|. \end{aligned}$$

Is it possible to extend this technique to derive a formula for four sets? □

The reader is encouraged to think about what the principle of inclusion-exclusion might say about the union of n finite sets, and compare with the results in Chapter 6.
As a final note, we remind the reader that the principle of inclusion-exclusion is often used in conjunction with the subtraction principle. In particular, if a large finite set contains n subsets, then we can apply the principle of inclusion-exclusion to the n subsets and then use complementary counting to find the number of elements outside of all of them but inside the overarching large set. This technique is repeatedly used in Section 6.2.

Chapter 2

Pigeonhole Principle

"... if you can get away with an argument as simple as the pigeon-hole principle, you have just been lucky."

– *Edsger Dijkstra, The Undeserved Status of the Pigeon-hole Principle*

'If the processes of abstraction and generalization... are of only limited use in combinatorics, then how can the subject be transmitted to future generations?... The important ideas of combinatorics do not usually appear in the form of precisely stated theorems, but more often as general principles of wide applicability."

– *Timothy Gowers, The Two Cultures of Mathematics*

The pigeonhole principle and its variations are deceptively simple statements that turn out to be surprisingly powerful and applicable tools in combinatorics. The central idea is that, when we are placing a finite non-zero number of pigeons into a finite non-zero number of holes, it is possible to assert the existence of a hole with a non-trivial lower bound on the number of pigeons in it, as well as a hole with an upper bound on the number of residing pigeons. No pigeons were harmed during the production of this text.

2.1 Injections and Surjections

Theorem 2.1 (Preimage principle)**.** Let X and Y be non-empty sets and $f : X \to Y$ be a function. Then:

1. For all $y, y' \in Y$, if $y \neq y'$ then the preimages $f^{-1}(y)$ and $f^{-1}(y')$ are disjoint.
2. If Y is a finite set $\{y_1, y_2, \dots, y_n\}$, then there exists a multiset
$$P = \langle f^{-1}(y_1), f^{-1}(y_2), \dots, f^{-1}(y_n)\rangle$$
of the preimages of all elements of Y. This multiset is a generalized partition of X (Definition 1.30).
3. If Y is finite, then all elements of the multiset P above are finite if and only if X is finite. If either case is assumed, then both X and all the elements of P are finite and
$$|X| = \sum_{y \in Y} |f^{-1}(y)|.$$

Proof. The statements are proven in sequence as each assertion assumes more than the one preceding it:

1. Since f is a function, each element of X gets mapped to at most one (in fact, exactly one) element of Y. If $f^{-1}(y)$ and $f^{-1}(y')$ shared a common element x, then x would get mapped to both of the distinct elements y and y', which is a contradiction.
2. Suppose Y is finite with n elements so that we can write it out as $Y = \{y_1, y_2, \dots, y_n\}$. Since Y is finite, there exists a multiset $P = \langle f^{-1}(y_1), f^{-1}(y_2), \dots, f^{-1}(y_n)\rangle$, and the above argument shows that P is a pairwise disjoint multiset of subsets of X. Note that we have prevented Y from being infinite because it might lead to infinitely many empty preimages, whereas the multiplicity of an element of a multiset must be a positive integer; it is for this reason that we have avoided writing P as $\langle f^{-1}(y) : y \in Y\rangle$, which is a piece of notation that might mislead the reader into believing the false notion that such a multiset can be constructed even in all cases of Y being infinite. To prove that P

is a generalized partition of X, all that remains to be shown is that the union of P is X. It suffices to show that every element of X is in some element of P, which is true because f being a function means each element of X gets mapped to at least one (again, exactly one) element of Y.

3. We continue to assume that Y is finite with n elements so that we may write it out as $Y = \{y_1, y_2, \ldots, y_n\}$. If all elements of P are finite, then P is a pairwise disjoint multiset of n finite sets. By the addition principle, their union X is finite. Conversely, if X is finite, then all of its subsets are finite. Since each element of P is a subset of X, they are all finite. Now suppose either hypothesis of this biconditional statement is true. Then P is an n-element generalized partition of a finite set X. By the addition principle (Theorem 1.31),

$$|X| = \sum_{i=1}^{n} |f^{-1}(y_i)| = \sum_{y \in Y} |f^{-1}(y)|.$$

□

Readers who are familiar with the axiom of choice might know of its controversial nature. In an effort to avoid using it in the ensuing material, we will instead apply the well-ordering principle in several places. This principle is stated as follows; it was mentioned in Volume 1 and it will be restated in Volume 3, after which it will be used several times in the context of number theory.

Theorem 2.2 (Well-ordering principle). Every non-empty set X of positive integers has a least element x_0. This means there exists an $x_0 \in X$ such that for all $x \in X$, if $x \neq x_0$ then $x_0 < x$.

Proof. The statement follows from the principle of mathematical induction (in fact, the two principles are equivalent), but it would be too much of a detour to state the proof here. □

Lemma 2.3 (Injection-surjection lemma). Let X and Y be non-empty sets such that Y is finite. Then:

1. There exists an injection $f : X \to Y$ if and only if there exists a surjection $g : Y \to X$.
2. If either side of the biconditional statement in the first part is assumed, then X is finite as well and $|X| \leq |Y|$.

Proof. There is a proof of the first assertion that does not rely on Y being finite, but the price of this generality is that one direction of the proof uses the axiom of choice in Volume 1. We will bypass the axiom of choice in the case that Y is finite by instead appealing to the well-ordering principle.

1. Suppose $f : X \to Y$ is an injection. By injectivity, for each $y \in Y$, the preimage $f^{-1}(y)$ has one or zero elements in it. Selecting some constant $z \in X$, we define $g : Y \to X$ so that $g(y)$ is the unique element of $f^{-1}(y)$ when $f^{-1}(y)$ is a singleton, and $g(y) = z$ when $f^{-1}(y)$ is empty. This is clearly a surjection onto X.

   Conversely, suppose $g : Y \to X$ is a surjection. The idea is to map each element $x \in X$ to some element of its preimage $g^{-1}(x)$ (the surjectivity of g implies that each such preimage is non-empty). By the preimage principle, if $y_1, y_2 \in Y$ such that $y_1 \neq y_2$, then the preimages $f^{-1}(y_1)$ and $f^{-1}(y_2)$ are disjoint, and so any function as described would be injective. The question is: by what source of power can this selection be made? The axiom of choice would allow us to arbitrary select an element of each preimage $g^{-1}(x)$, but we have already stated that this tool is unnecessarily strong. What we do is fix a bijection $h : Y \to [n]$ where $n = |Y|$. Then for each $x \in X$, we use the well-ordering principle to choose from among the elements of $g^{-1}(x)$ the element $y \in Y$ with the smallest image under h. This is how we define an injection $f : X \to Y$.

2. Suppose there exists an injection $f : X \to Y$. Let the range of f be $Z = \{f(x) : x \in X\}$. Then $f : X \to Z$ is a bijection because every function is surjective onto its range, and $f : X \to Z$ inherits the injectivity of $f : X \to Y$. Since $Z \subseteq Y$ and Y is finite, Z is finite. As $f : X \to Z$ is a function, Z cannot be empty and so there exists a positive integer n such that $Z \approx [n]$. Combined with $X \approx Z$, this means $X \approx [n]$ and so X is finite. Finally, we need to show that $|X| \leq |Y|$. This is true because,

by the bijection principle, $X \approx Z$ implies $|X| = |Z|$, and since $Z \subseteq Y$, we know that $|Z| \leq |Y|$ (see Problem 1.9). Thus, $|X| = |Z| \leq |Y|$.

We could have also started by assuming that there exists a surjection $g : Y \to X$, and the previous part would imply the existence of an injection $f : X \to Y$ so the argument would be the same.

□

Problem 2.4. Suppose X and Y are non-empty finite sets. Show that there exists an injection $f : X \to Y$ or an injection $f : Y \to X$. As a consequence, the injection-surjection lemma tells us that, in whichever direction an injection exists, a surjection exists in the other direction.

Note that the injection-surjection lemma is more general than the upcoming pigeonhole principle (Theorem 2.5) and its reverse (Theorem 2.6) because the former proves that X is finite, whereas the latter two assume it. However, the strong versions (Theorem 2.12 and Theorem 2.14) arguably assert even more than the injection-surjection lemma.

Theorem 2.5 (Pigeonhole principle)**.** Suppose X and Y are non-empty finite sets. If $|X| > |Y|$, then any function $f : X \to Y$ is not injective. Informally, this means that if we are placing pigeons into holes and there are more pigeons than holes, then some hole will receive two or more pigeons.

Proof. Let X and Y be non-empty finite sets. The contrapositive of the stated assertion is: If there exists an injection $f : X \to Y$ then $|X| \leq |Y|$. This is immediate from the injection-surjection lemma. □

Theorem 2.6 (Reverse pigeonhole principle)**.** Suppose X and Y are non-empty finite sets. If $|X| < |Y|$ then any function $f : X \to Y$ is not surjective. Informally, this means that if we are placing pigeons into holes and there are fewer pigeons than holes, then some hole will be empty.

Proof. Let X and Y be non-empty finite sets. The contrapositive of the stated assertion is: If there exists a surjection $f : X \to Y$ then $|X| \geq |Y|$. Again, this is immediate from the injection-surjection lemma. □

As with all pigeonhole-type theorems, we advise the reader to notice that the pigeonhole principle and its reverse are not *constructive* theorems that tell us exactly where a special occurrence takes place. Rather, they are *existential* theorems. The location of existence cannot be specified or made even moderately more precise without more information.

Problem 2.7. Suppose X and Y are non-empty finite sets such that $|X| = |Y|$, and that $f : X \to Y$ is a function. Show that f is an injection if and only if f is a surjection; assuming either case, it means f is a bijection. Informally, we are saying that if some pigeons are placed into just as many many holes, then each hole receives at most one pigeon if and only if it receives at least one pigeon. In Definition 3.6, we will define a permutation of a non-empty finite set S of cardinality n to be an injection $f : [n] \to S$. As a corollary, show that this means that if S is a non-empty finite set of cardinality n, then a function $f : [n] \to S$ is a permutation of S if and only if $f : [n] \to S$ is a bijection.

The following is the finite case of a powerful result from set theory for establishing the existence of a bijection.

Theorem 2.8 (Finite Schröder-Bernstein theorem)**.** Suppose X and Y are non-empty sets such that at least one is finite. If there exists an injection $f : X \to Y$ and an injection $g : Y \to X$, then X and Y are both finite and $|X| = |Y|$. As a consequence, we conclude that $X \approx Y$.

Proof. Recall the following result from our injection-surjection lemma (Lemma 2.3): If A and B are non-empty sets such that B is finite and there exists an injection $h : A \to B$, then A is finite as well and $|A| \leq |B|$. We will use this result two times.

Suppose X and Y are non-empty sets. If Y is finite, then the existence of the injection $f : X \to Y$ implies that X is finite. Similarly, if X is finite, then the existence of the injection $g : Y \to X$ implies that Y is finite. Since we know that at least one of X, Y is finite, this means both are finite. By the injectivity of f, we get $|X| \leq |Y|$. Similarly, by the injectivity of g, we get $|Y| \leq |X|$. Thus, $|X| = |Y|$. By the bijection principle, $X \approx Y$. □

Problem 2.9. Let X and Y be non-empty sets such that at least one is finite. In each of the following cases, show that X and Y are both finite and $|X| = |Y|$, implying $X \approx Y$.

1. There exists a surjection $f : X \to Y$ and a surjection $g : Y \to X$.
2. There exists a surjection $f : X \to Y$ and an injection $h : X \to Y$.

These are variants of the finite Schröder-Bernstein theorem that are occasionally useful.

We now offer a novel application of the pigeonhole principle, as an example of its power.

Example 2.10. Let α be an irrational number. Show that the set of fractional parts $S = \{\{i\alpha\} : i \in \mathbb{Z}\}$ is "dense in the the reals modulo 1," meaning every pair of distinct elements of $[0, 1)$ contains a member of S strictly in between the two elements.

Solution. Let α be a fixed irrational number. We first claim that for different integers i, j, $\{i\alpha\}$ and $\{j\alpha\}$ cannot differ by a rational amount. In particular, the multiset of fractional parts $\{i\alpha\}$, while i ranges over the integers, has all distinct elements, since no pair of elements can differ by the rational number 0. The proof of the claim is as follows. Suppose, for contradiction, that there exists a rational q such that

$$\{j\alpha\} - \{i\alpha\} = q$$

for $i \neq j$. Using the definition that $\{x\} = x - \lfloor x \rfloor$ for any real number x,

$$\begin{aligned} q &= \{j\alpha\} - \{i\alpha\} = (j\alpha - \lfloor j\alpha \rfloor) - (i\alpha - \lfloor i\alpha \rfloor) = (j - i)\alpha - (\lfloor j\alpha \rfloor - \lfloor i\alpha \rfloor) \\ \alpha &= \frac{q + (\lfloor j\alpha \rfloor - \lfloor i\alpha \rfloor)}{j - i} \in \mathbb{Q}. \end{aligned}$$

This contradicts the irrationality of α.
Next comes the part where pigeonhole is relevant. We claim that, for any positive integer n, if we split $[0, 1)$ into n disjoint intervals of equal length $\frac{1}{n}$, as in

$$[0, 1) = \left[0, \frac{1}{n}\right) \sqcup \left[\frac{1}{n}, \frac{2}{n}\right) \sqcup \cdots \sqcup \left[\frac{n-1}{n}, \frac{n}{n}\right),$$

then there exists some such interval in this split with at least two elements of S. This is easily true by the pigeonhole principle because we showed that S has infinitely many distinct elements, and there are only finitely many intervals in the disjoint union. So, there exist integers i, j and some $k \in [n - 1]^* = \{0, 1, 2, \ldots, n - 1\}$ such that

$$\frac{k}{n} \leq \{i\alpha\} < \{j\alpha\} < \frac{k + 1}{n}.$$

Manipulating these inequalities, we get

$$0 < \{j\alpha\} - \{i\alpha\} < \frac{1}{n}.$$

So, the set of positive integers whose reciprocals are greater than $\{j\alpha\} - \{i\alpha\}$ is non-empty (moreover, no such reciprocal, which is rational, can equal $\{j\alpha\} - \{i\alpha\}$ by our first argument above). It is interesting to consider such integers k because then

$$\begin{aligned} & \{j\alpha\} - \{i\alpha\} < \frac{1}{k} \\ \implies & k(\{j\alpha\} - \{i\alpha\}) < 1 \\ \implies & k(\{j\alpha\} - \{i\alpha\}) = \{k(\{j\alpha\} - \{i\alpha\})\} = \{k((j\alpha - \lfloor j\alpha \rfloor) - (i\alpha - \lfloor i\alpha \rfloor))\} \\ & \qquad = \{k(j - i)\alpha - k(\lfloor j\alpha \rfloor - \lfloor i\alpha \rfloor)\} = \{k(j - i)\alpha\} \in S, \end{aligned}$$

since $k(\lfloor j\alpha \rfloor - \lfloor i\alpha \rfloor) \in \mathbb{Z}$. So, up to a point, the positive integer multiples of $\beta = \{j\alpha\} - \{i\alpha\}$ are elements of S. Let t be the largest positive integer such that $t\beta < 1$ so that $1 < (t + 1)\beta$ (note that equality cannot hold in either inequality, as stated earlier). Since each interval $\left[\frac{m}{n}, \frac{m+1}{n}\right)$ for $m \in [n-1]^*$ has length $\frac{1}{n}$, and the distance between $\ell\beta$ and $(\ell+1)\beta$, for any positive integer ℓ, is β, which satisfies $0 < \beta < \frac{1}{n}$, we find that $k\beta \in \left[\frac{m}{n}, \frac{m+1}{n}\right)$ for some integer $k \in [t] = \{1, 2, \ldots, t\}$ (imagine tiling). Since n can be arbitrarily large, every non-trivial interval contains an interval $\left[\frac{m}{n}, \frac{m+1}{n}\right)$ for some n and m, which in turn we have shown to contain some element of S. □

The following is an interesting problem, for those who are familiar with modular arithmetic.

Problem 2.11 (Thue's lemma). Let p be a prime and a be an integer such that $p \nmid a$. Prove that the linear congruence

$$ax \equiv y \pmod{p}$$

has a solution $(x, y) = (x_0, y_0)$ such that

$$0 < |x_0| < \sqrt{p} \text{ and } 0 < |y_0| < \sqrt{p}.$$

2.2 Extending Pigeonhole

In the way that it was stated, the pigeonhole principle and its reverse are a bit weak. Now we will provide stronger statements.

Theorem 2.12 (Strong pigeonhole principle). Let X and Y be non-empty finite sets and $f : X \to Y$ be a function. Then there exists an element $y \in Y$ such that

$$|f^{-1}(y)| \geq \left\lceil \frac{|X|}{|Y|} \right\rceil.$$

This has the following implications of decreasing generality. Let n and k be positive integers.

1. If n pigeons are placed into k holes, then there exists a hole with at least $\left\lceil \frac{n}{k} \right\rceil$ pigeons.
2. If $k(n-1)+1$ pigeons are placed into k holes, then there exists a hole with at least n pigeons.
3. If $k+1$ pigeons are placed into k holes, then there exists a hole with at least 2 pigeons.

Proof. We will prove the formal statement and leave the implications to the reader. Let X, Y and $f : X \to Y$ be as stated. Suppose, for contradiction, that, for all $y \in Y$,

$$|f^{-1}(y)| < \left\lceil \frac{|X|}{|Y|} \right\rceil.$$

Using the fact from Volume 1 that $\lceil x \rceil < x + 1$ for all real x, and since both sides of the above inequality are integers, we can convert to a non-strict inequality and get

$$|f^{-1}(y)| \leq \left\lceil \frac{|X|}{|Y|} \right\rceil - 1 < \left(\frac{|X|}{|Y|} + 1\right) - 1 = \frac{|X|}{|Y|}$$

By the preimage principle,

$$|X| = \sum_{y \in Y} |f^{-1}(y)| < \sum_{y \in Y} \frac{|X|}{|Y|} = |Y| \cdot \frac{|X|}{|Y|} = |X|,$$

which is a contradiction. The desired conclusion follows. □

Problem 2.13. Show that the strong pigeonhole principle implies the pigeonhole principle (Theorem 2.5).

Theorem 2.14 (Strong reverse pigeonhole principle)**.** Let X and Y be non-empty finite sets and $f : X \to Y$ be a function. Then there exists an element $y \in Y$ such that

$$|f^{-1}(y)| \leq \left\lfloor \frac{|X|}{|Y|} \right\rfloor .$$

This has the following implications of decreasing generality. Let n and k be positive integers.

1. If n pigeons are placed into k holes then there exists a hole with at most $\left\lfloor \frac{n}{k} \right\rfloor$ pigeons.
2. If $k(n+1)-1$ pigeons are placed into k holes, then there exists a hole with at most n pigeons.
3. If $2k-1$ pigeons are placed into k holes, then there exists a hole with at most 1 pigeon.

Proof. As with the strong pigeonhole principle, we will only prove the most general result. Let X, Y and $f : X \to Y$ be as stated. Suppose, for contradiction, that for all $y \in Y$,

$$|f^{-1}(y)| > \left\lfloor \frac{|X|}{|Y|} \right\rfloor .$$

Using the fact from Volume 1 that $\lfloor x \rfloor > x - 1$ for all real x, and since both sides of the above inequality are integers, we can convert to a non-strict inequality and get

$$|f^{-1}(y)| \geq \left\lfloor \frac{|X|}{|Y|} \right\rfloor + 1 > \left(\frac{|X|}{|Y|} - 1 \right) + 1 = \frac{|X|}{|Y|}.$$

By the preimage principle,

$$|X| = \sum_{y \in Y} |f^{-1}(y)| > \sum_{y \in Y} \frac{|X|}{|Y|} = |Y| \cdot \frac{|X|}{|Y|} = |X|,$$

which is a contradiction. Thus, our initial assumption was wrong and its negation is true. □

Problem 2.15. Show that the strong reverse pigeonhole principle implies the reverse pigeonhole principle (Theorem 2.6).

Theorem 2.16. For each possible pair of non-empty finite sets X and Y, the lower bound $\left\lceil \frac{|X|}{|Y|} \right\rceil$ in the strong pigeonhole principle is the highest possible one that we can guarantee, and the upper bound $\left\lfloor \frac{|X|}{|Y|} \right\rfloor$ in the strong reverse pigeonhole principle is the lowest possible one. In other words, given non-empty finite sets X and Y:

1. There exists a function $f : X \to Y$ such that the preimage of each $y \in Y$ has less than or equal to $\left\lceil \frac{|X|}{|Y|} \right\rceil$ elements.
2. There exists a function $g : X \to Y$ such that the preimage of each $y \in Y$ has greater than or equal to $\left\lfloor \frac{|X|}{|Y|} \right\rfloor$ elements.

This shows the optimality of our results, if there is no additional information available about f.

Proof. Let X and Y be non-empty finite sets.

1. We start by placing disjoints sets of $\left\lceil \frac{|X|}{|Y|} \right\rceil - 1$ balls from X into each of the boxes in Y. To ensure that this is possible, we will check that there are enough balls, meaning

$$0 < |X| - \left(\left\lceil \frac{|X|}{|Y|} \right\rceil - 1 \right) \cdot |Y|.$$

It would also be nice if it happened to be true that

$$|X| - \left(\left\lceil \frac{|X|}{|Y|} \right\rceil - 1 \right) \cdot |Y| \leq |Y|,$$

because then we could distribute the remaining balls across the $|Y|$ boxes in any way that each box gets at most one more ball (some might get none of these extras), which would complete the construction of f. Rearranging the double inequality

$$0 < |X| - \left(\left\lceil \frac{|X|}{|Y|} \right\rceil - 1 \right) \cdot |Y| \leq |Y|,$$

we find that it is equivalent to

$$\left\lceil \frac{|X|}{|Y|} \right\rceil - 1 < \frac{|X|}{|Y|} \leq \left\lceil \frac{|X|}{|Y|} \right\rceil .$$

This is true because for all real numbers x, $\lceil x \rceil = n$ if and only if $n - 1 < x \leq n$.

2. We start by placing disjoint sets of $\left\lfloor \frac{|X|}{|Y|} \right\rfloor$ balls into each box. To check that this is possible, we will show that there are enough elements, meaning

$$0 \leq |X| - \left\lfloor \frac{|X|}{|Y|} \right\rfloor \cdot |Y|.$$

This is true because it is equivalent to $\left\lfloor \frac{|X|}{|Y|} \right\rfloor \leq \frac{|X|}{|Y|}$, and the floor function of a real number never exceeds the number. Any remaining balls can be distributed among the boxes in any way and this completes the construction of g.

□

Problem 2.17. Let $n_1, n_2, \ldots, n_k$ be $k \geq 1$ positive integers. Prove that:

1. If $n_1 + n_2 + \cdots + n_k - k + 1$ pigeons are placed into k holes that are ordered from 1 to k inclusive, then there exists an index $1 \leq i \leq k$ such that hole number i receives at least n_i pigeons.

2. If $n_1 + n_2 + \cdots + n_k + k - 1$ pigeons are placed into k holes that are ordered from 1 to k inclusive, then there exists an index $1 \leq i \leq k$ such that hole number i receives at most n_i pigeons.

We have stated this problem informally because, due to ordered nature of the holes, it is cumbersome to express it in terms of sets and functions.

We will again apply the pigeonhole principle and pigeonhole-type ideas to interesting problems when we study Ramsey theory in Section 8.3.

Chapter 3

Multiplication and Division

"First, it is necessary to study the facts, to multiply the number of observations, and then later to search for formulas that connect them so as thus to discern the particular laws governing a certain class of phenomena. In general, it is not until after these particular laws have been established that one can expect to discover and articulate the more general laws that complete theories by bringing a multitude of apparently very diverse phenomena together under a single governing principle."

– *Augustin-Louis Cauchy*

"The key issue for me is finding the right definitions, finding the right notions that really capture the essence of some mathematical phenomenon. I often have some vague vision of what I want to understand, but I'm often missing the words to really say that... suddenly it clicks, and suddenly I can say what I always wanted to say."

– *Peter Scholze, ICM 2018*

Our next steps are combinatorial multiplication and division. Multiplication is about counting Cartesian products of finite sets or certain special subsets of such products. Division is about splitting a set into equal parts. We will see a weak and a strong version of each of multiplication and division. These principles will allow us to develop formulas for the three most ubiquitous combinatorial expressions: permutations, combinations or binomial coefficients, and multinomial coefficients.

3.1 Multiplication Principle

Some sources refer to the sets A_i in the Cartesian product (see Volume 1)

$$A_1 \times A_2 \times \cdots \times A_n$$

as being "independent" from each other because the entry at each index i has full freedom to be any element of A_i without being hindered by the selection of entries at any of the other indices. This inspired the name of the following principle.

Theorem 3.1 (Independent multiplication principle)**.** Suppose n is a positive integer. For any list of non-empty sets $(A_1, A_2, \ldots, A_n)$, all of the A_i are finite if and only if $A_1 \times A_2 \times \cdots \times A_n$ is finite. If either case is assumed, then all of the A_i and their Cartesian product are finite, and

$$|A_1 \times A_2 \times \cdots \times A_n| = |A_1| \cdot |A_2| \cdots |A_n|.$$

This formula actually also works if some of the A_i are empty because then the Cartesian product would also be empty, which means that it is true that both sides of the above cardinality equation are 0.

Proof. Before writing a formal proof, we note that, for two or three sets, the formula makes intuitive sense because the Cartesian product of two finite sets with p and q elements can be visually represented as the squares resulting from splitting a $p \times q$ rectangle into pq unit squares. For three sets, we can go to three dimensions and split a $p \times q \times r$ rectangular prism into pqr unit cubes. Now we tackle it formally.
For one direction, if $A_1 \times A_2 \times \cdots \times A_n$ is finite, then for each index $1 \leq i \leq n$, the map (called a projection)

$$\begin{aligned} f_i : A_1 \times A_2 \times \cdots \times A_n &\to A_i \\ (a_1, a_2, \ldots, a_n) &\mapsto a_i \end{aligned}$$

is a surjection, which proves that A_i is finite, by the injection-surjection lemma.
Conversely, suppose all of the A_i are finite. We will prove by induction on $n \geq 1$ that $A_1 \times A_2 \times \cdots \times A_n$ is finite and that

$$|A_1 \times A_2 \times \cdots \times A_n| = |A_1| \cdot |A_2| \cdots |A_n|.$$

This will complete the proof because, if the assumption were instead that $A_1 \times A_2 \times \cdots \times A_n$ is finite then the first argument would imply that all of the A_i are finite, at which point we could invoke the second argument (although the part of the second argument proving that $A_1 \times A_2 \times \cdots \times A_n$ is finite would be redundant in this case).
We proceed by induction on $n \geq 1$. In the base case $n = 1$, the assertion is that $|A_1| = |A_1|$, which is obviously true. Now suppose the proposition is true for some integer $n \geq 1$. Let

$$(A_1, A_2, \ldots, A_n, A_{n+1})$$

be a list of $n + 1$ non-empty finite sets. The idea is to partition $A_1 \times A_2 \times \cdots \times A_n \times A_{n+1}$ into cases according to the entry selected from the rightmost set A_{n+1}, though it is easier to implement with the preimage principle than using the addition principle. We define the function

$$\begin{aligned} f : A_1 \times A_2 \times \cdots \times A_n \times A_{n+1} &\to A_1 \times A_2 \times \cdots \times A_n \\ (a_1, a_2, \ldots, a_n, a_{n+1}) &\mapsto (a_1, a_2, \ldots, a_n), \end{aligned}$$

which simply drops the entry at the last index. By the induction hypothesis, $A_1 \times A_2 \times \cdots \times A_n$ is finite and

$$|A_1 \times A_2 \times \cdots \times A_n| = |A_1| \cdot |A_2| \cdots |A_n|.$$

Note that

$$f^{-1}((a_1, a_2, \ldots, a_n)) = \{(a_1, a_2, \ldots, a_n, a) : a \in A_{n+1}\} \approx A_{n+1}.$$

Due to this equipotence and the bijection principle, the finite nature of A_{n+1} implies that the preimage of every element of $A_1 \times A_2 \times \cdots \times A_n$ is finite with cardinality $|A_{n+1}|$. By the preimage principle, this means that $A_1 \times A_2 \times \cdots \times A_n \times A_{n+1}$ is finite and

$$\begin{aligned} |A_1 \times A_2 \times \cdots \times A_n \times A_{n+1}| &= \sum_{\substack{(a_1,a_2,\ldots,a_n) \\ \in A_1 \times A_2 \times \cdots \times A_n}} |f^{-1}((a_1, a_2, \ldots, a_n))| \\ &= \sum_{\substack{(a_1,a_2,\ldots,a_n) \\ \in A_1 \times A_2 \times \cdots \times A_n}} |A_{n+1}| \\ &= |A_1 \times A_2 \times \cdots \times A_n| \cdot |A_{n+1}| \\ &= |A_1| \cdot |A_2| \cdots |A_n| \cdot |A_{n+1}|. \end{aligned}$$

This completes the induction. □

Example 3.2. Suppose n is a non-negative integer. If S is a finite set with n elements, show that $\mathcal{P}(S)$ is finite and determine its cardinality. Power sets were defined in Volume 1.

Solution. If S is empty, then the only element in $\mathcal{P}(S)$ is $\varnothing$, so $|\mathcal{P}(\varnothing)| = 1$. Now we can assume that $n \geq 1$. We will highlight an approach that uses the independent multiplication principle. First, we arbitrarily label the elements of S as $\{a_1, a_2, \ldots, a_n\}$ so that we can refer to them in terms of an order; formally, we are formatting S as a list. Then we can represent each set $T \in \mathcal{P}(S)$ as an n-tuple of 0's and 1's so that, for each index $1 \leq i \leq n$, the entry at index i is

$$\begin{cases} 0 & \text{if } a_i \notin T \\ 1 & \text{if } a_i \in T \end{cases}.$$

Let $\{0, 1\}^n$ be the Cartesian product of n copies of the set $\{0, 1\}$, so $\{0, 1\}^n$ is the set of all n-tuples whose entries consist of only 0's and 1's. By the independent multiplication principle, $\{0, 1\}$ being finite implies that $\{0, 1\}^n$ is finite, and

$$|\{0, 1\}^n| = |\{0, 1\}|^n = 2^n.$$

We leave it to the reader to prove that $\{0,1\}^n \approx \mathcal{P}(S)$ using the stated correspondence. By the bijection principle, $\mathcal{P}(S)$ is finite and

$$|\mathcal{P}(S)| = |\{0,1\}^n| = 2^n.$$

This is consistent with there being $2^0 = 1$ elements in the power set of the empty set. □

Problem 3.3. Write out all of the elements of the power set of $\{a, b, c, d\}$.

Problem 3.4. There are two blue cubes labelled $1, 2$, three green cubes labelled a, b, c and four red cubes labelled $\alpha, \beta, \gamma, \delta$. In how many ways can we select one blue cube, one green cube, and one red cube?

Problem 3.5. Let k be an positive integer. Determine the number of k-tuples

$$(a_1, a_2, \ldots, a_k) \in \{-1, 1\}^k$$

such that

$$a_1 a_2 \cdots a_k = 1.$$

Given a finite set, we might wish to construct a list that contains only elements of that set without repetition (though not necessarily every element of the set would appear in the list). Using the language of sets and functions, we can formally define such an object as follows.

Definition 3.6. If S is a non-empty finite set of cardinality n and k is an integer such that $1 \leq k \leq n$, then a ***k*-permutation** of S is an injective function $f : [k] \to S$. If $k = n$, then we drop the prefix and refer to an n-permutation as a **permutation** of S.

Much of combinatorics is about permutations of finite sets, as will become evident to the reader through experience.
A natural question whose answer the curious mind might seek is: How many permutations of a non-empty finite S exist? Unfortunately, the set of permutations of S is not the Cartesian product of the set of possible entries at each index, as that would be S^n. There is far too much freedom in S^n, whereas the set of permutations of S should satisfy the fact that fixing the entry at just one index prevents the entry at any other index from assuming the same value. Thus, there is no obvious way to apply the independent multiplication principle to the set of permutations of S, which is a proper subset of S^n. We need a stronger idea that applies to suitable subsets of a Cartesian product and not only the full Cartesian product.

Definition 3.7. A temporary term that we will be using that we must clarify early is "valid." Suppose T is a set of lists that all have the same length. If we say that the certain indices have been given a **valid assignment** of entries, we mean that the entries at those indices have been fixed in a way that there is at least one way to fill the entries at the remaining indices to produce a list from T. Similarly, if certain initial indices (possibly no indices) have been given a valid assignment of entries, then the **valid entries** at an empty available index are the elements with which the entry at that index can be fixed so that the initially fixed entries along with the newly fixed entry together form a valid assignment of entries for the initial indices and this new index. We have made this tongue-twisting clarification so that it is not misunderstood that we are ever filling entries in a way that does not produce an element of T.

Our visualization of a Cartesian product of two or three non-empty finite sets was a rectangle or rectangular prism. A way to visualize a general collection of n-tuples is as a tree diagram. This is not the place for a formal definition of a graph-theoretic tree (for that, see Definition 8.7), but the basic idea is to:

0. Step 0: Start off with a general node called a root.
1. Step 1: From the root, branch out to new nodes, once for each valid entry at the first index.
2. Steps $k \geq 2$: For each of the nodes created at step $k - 1$, branch out to new nodes, once for each distinct valid entry at the kth index such that the first $k - 1$ entries of the corresponding permutation are those that can be traced back in an ascending manner through the tree.

This process continues until the completion of step n, which is the length of the tuples.
The following is the tree of permutations of $\{1, 2, 3\}$.

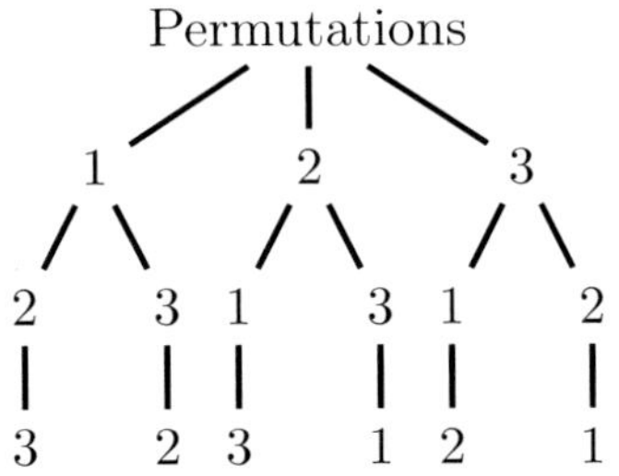

A property that is evident in this tree is that, for $k = 1, 2$, no matter how the first k indices are given a valid assignment of entries, the number of subsequent valid entries at the $(k+1)^{\text{th}}$ index is always the same, even though different valid assignments of entries to the first k indices do not lead to the same set of valid entries at the $(k+1)^{\text{th}}$ index. This is a regularity that often occurs in combinatorics and it is the key to a new multiplication principle.

Definition 3.8. Let n be a positive integer and let T be a non-empty set of n-tuples, say $(x_1, x_2, \dots, x_n)$. Suppose it is true that: for each index k such that $2 \leq k \leq n$, there exists a positive integer d_k such that, for every fixed k-tuple $(a_1, \dots, a_{k-1})$, if the set

$$T(a_1, \dots, a_{k-1}) = \{(x_1, x_2, \dots, x_n) \in T : x_1 = a_1, x_2 = a_2, \dots, x_{k-1} = a_{k-1}\}$$

is non-empty, then the set of possible entries x_k at the kth index among all elements of $T(a_1, \dots, a_{k-1})$ is finite with cardinality d_k. In the language of Definition 3.7, for every valid assignment of entries to the first $k-1$ indices, the number of valid entries at index k is d_k. Note that d_k depends only on k and not on any particular choice of the first $k-1$ entries $(a_1, \dots, a_{k-1})$. Then we call T **symmetrically dependent**. Letting d_1 be the number of valid entries at the first index x_1 among all elements of T and without fixing any other entry, we call d_k the kth **dependence number** for each index $1 \leq k \leq n$.

Theorem 3.9 (Dependent multiplication principle)**.** *Suppose n is a positive integer. Let T be a symmetrically dependent set of n-tuples, whose k^{th} dependence number is d_k for each index $1 \leq k \leq n$. Then T is finite with cardinality*

$$|T| = d_1 d_2 \cdots d_n.$$

Proof. The proof is by induction on $n \geq 1$. In the base case $n = 1$, it is clearly true that T is finite with $|T| = d_1$ because, if T contains only 1-tuples, then the set of elements in T is just the set of valid entries at the first index. Now, suppose the assertion is true for some integer $n \geq 1$. Let T be a symmetrically dependent set of $(n+1)$-tuples with dependence numbers $d_1, d_2, \dots, d_n, d_{n+1}$. Inspired by the proof of the independent multiplication principle, we try dropping the entry at the last index. Let

$$R = \{(a_1, a_2, \dots, a_n) : \exists a, (a_1, a_2, \dots, a_n, a) \in T\}.$$

So R is the set of all valid assignments of entries to the first n indices of elements of T, where each valid assignment is written as an n-tuple. It is not difficult to see that R inherits the symmetrically dependent status of T with dependence numbers $d_1, d_2, \dots, d_n$. For those unconvinced, look at a tree diagram for inspiration: the tree of R just shears off the "leaves" of the tree of T. In order to invoke the preimage principle, we define the projection map

$$\begin{aligned} f : T &\to R \\ (a_1, a_2, \dots, a_n, a_{n+1}) &\mapsto (a_1, a_2, \dots, a_n), \end{aligned}$$

which simply makes the last entry disappear. By the induction hypothesis, R is finite with cardinality $|R| = d_1 d_2 \cdots d_n$. Note that

$$\begin{aligned} f^{-1}((a_1, a_2, \dots, a_n)) &= \{(x_1, x_2, \dots, x_n, a) \in T : x_1 = a_1, x_2 = a_2, \dots, x_n = a_n\} \\ &\approx \{a : (a_1, a_2, \dots, a_n, a) \in T\}, \end{aligned}$$

where the last set is the set of valid entries at index $n+1$, given that the entries at the first n indices are fixed as $(a_1, a_2, \dots, a_n)$. Since T is symmetrically dependent, $\{a : (a_1, a_2, \dots, a_n, a) \in T\}$ is finite with cardinality

d_{n+1} for any choice of $(a_1, a_2, \ldots, a_n) \in R$. Thanks to the above equipotence, the bijection principle says that all the preimages $f^{-1}((a_1, a_2, \ldots, a_n))$ are also finite with cardinality d_{n+1}. By the preimage principle, this means that T is finite with cardinality

$$|T| = \sum_{(a_1,a_2,\ldots,a_n)\in R} |f^{-1}((a_1, a_2, \ldots, a_n))|$$
$$= \sum_{(a_1,a_2,\ldots,a_n)\in R} d_{n+1} = |R| \cdot d_{n+1} = d_1 d_2 \cdots d_n d_{n+1}.$$

This completes the induction. □

In our experience, numerous sources do not fully state the dependent multiplication principle and its proof, or make the mistake of applying the dependent one as if it is a trivial consequence of the independent one (as far as we know, this deduction is not possible). For ease of language, now that we have proven both, we might refer to either as "the multiplication principle."

Problem 3.10. Show that the Cartesian product of a non-empty list of non-empty finite sets is symmetrically dependent. As such, the independent multiplication principle is a special case of the dependent multiplication principle.

Definition 3.11. For each positive integer n, the **factorial** of n is denoted by and defined as

$$n! = n(n-1)\cdots 2 \cdot 1.$$

The factorial of 0 is defined as $0! = 1$, which is a definition that will make sense in the context of the formula for the number of permutations of a set (Theorem 3.12). Alternatively, $0!$ can be considered to be the empty product, which takes the value of the multiplicative identity 1.

Theorem 3.12. Let n and k be integers such that $1 \le k \le n$. If S is a finite set with n elements, then the set of k-permutations of S is finite and has cardinality

$$n(n-1)\cdots(n-k+1) = \frac{n!}{(n-k)!}.$$

As a consequence, the number of permutations of S is

$$n(n-1)\cdots 2 \cdot 1 = n!.$$

Proof. We will apply the dependent multiplication principle. It boils down to proving that the set of k-permutations of S is symmetrically dependent and determining the dependence numbers d_i. Any of the elements of S is a valid entry for the first index of a k-permutation, so d_1 exists and is equal to n. For any index i such that $2 \le i \le n$, fixing the entries of the first $i-1$ indices of a k-permutation in any valid way implies that $i-1$ distinct elements of S have been "used up," and so there are $n-(i-1) = n-i+1$ remaining valid entries for the ith index. This proves that, for each index $2 \le i \le n$, the ith dependence number d_i exists and is equal to $n-i+1$. Therefore, we can invoke the dependent multiplication principle to get the answer

$$d_1 d_2 \cdots d_k = n(n-1)\cdots(n-k+1) = \frac{n!}{(n-k)!}.$$

For $k = n$, this is $n!$. □

It is now evident that our definition $0! = 1$ is compatible with saying that the number of 0-permutations of any finite set, including the empty set, is $\frac{n!}{(n-0)!} = 1$. This matches our intuition that there is 1 way of permuting nothing, which is to do nothing. Apparently, doing nothing is doing something.

Definition 3.13. Let S be a (possibly empty) finite set of cardinality n and let k be a non-negative integer such that $0 \le k \le n$. If k and n satisfy $1 \le k \le n$, then we keep our existing definition of k-permutations of S. As an extension, we define that the only 0-permutation of S is the empty list (), which we can justify in retrospect by saying that there is only one way to order zero elements. In general, we will let $P(n,k)$ denote the number of k-permutations of a set of n elements, where k and n are any integers that satisfy $0 \le k \le n$. If either of k or n is 0 (note that $n = 0$ automatically implies $k = 0$), then our definition implies that $P(n,k) = 1$. In all cases,

$$P(n,k) = \frac{n!}{(n-k)!}.$$

Problem 3.14. If S is a non-empty finite set, prove that the set of bijections $b : S \to S$ is finite and determine its cardinality in terms of $|S| = n$. This problem is somewhat technical, but solving it by using a bijection will give the reader good practice in mapping functions to functions.

Problem 3.15. Let n be positive integer. There are n people who must be seated in a row. Out of the n people, m of them are friends who want to sit in consecutive seats, where $1 \le m \le n$. In how many ways can the n people be seated?

Problem 3.16. Let n be a positive integer. Find the number of ways in which n couples can sit on a bench with $2n$ seats so that each couple is sitting together.

3.2 Correspondence Principle

Theorem 3.17 (Division principle)**.** Suppose A is a non-empty set, and n and k are positive integers. If $\{A_1, A_2, \ldots, A_n\}$ is a partition of A such that each A_i is finite and

$$|A_1| = |A_2| = \cdots = |A_n| = k,$$

then A is finite and $n = \dfrac{|A|}{k}$.

Proof. By the addition principle, $A = A_1 \cup A_2 \cup \cdots \cup A_n$ is finite and

$$|A| = |A_1| + |A_2| + \cdots + |A_n| = \underbrace{k + k + \cdots + k}_{n \text{ copies of } k} = n \cdot k.$$

Thus, we can isolate n to get $n = \dfrac{|A|}{k}$. □

The way that it has been stated, the above division principle (Theorem 3.17) is a side effect of the addition principle that is not very useful. However, it inspires a more sophisticated notion of combinatorial division, which we will see in Theorem 3.19.

Definition 3.18. Suppose X and Y are non-empty sets, and that $f : X \to Y$ is a function. If there exists a positive integer k such that the preimage of every $y \in Y$ is a finite set with cardinality k, then f is said to be a k**-to-1 correspondence**.

Example. Every k-to-1 correspondence is a surjection, but a surjection is not necessarily a k-to-1 correspondence. Moreover, a function is a 1-to-1 correspondence if and only if it is a bijection; this is proven in the solution to Problem 3.20.

Theorem 3.19 (Correspondence principle)**.** Suppose X and Y are non-empty sets such that at least one of them is finite, and let k be a positive integer. Then there exists a k-to-1 correspondence $f : X \to Y$ if and only if both X, Y are finite and $|Y| = \dfrac{|X|}{k}$.

Proof. Let X and Y be non-empty finite sets such that at least one of them is finite (we will do casework on which one is finite). For one direction, let k be a positive integer and $f : X \to Y$ be a k-to-1 correspondence. There are two possibilities:

- Suppose X is finite. Since f is a k-to-1 correspondence, f is a surjection. By the injection-surjection lemma, Y is finite as well.
- Suppose Y is finite. Since the preimage of each element of Y is finite, the preimage principle says that the domain X is finite.

If either case is assumed, then both X and Y are finite, and the preimage principle asserts that

$$|X| = \sum_{y \in Y} |f^{-1}(y)| = \sum_{y \in Y} k = |Y| \cdot k.$$

Our proof of the other direction will make more sense to those who have some familiarity with the equivalence classes of modular arithmetic, but we have written it in an accessible way. We assume that both X, Y are finite and $k \cdot |Y| = |X|$. We want to construct a k-to-1 correspondence $f : X \to Y$. Since X and Y are finite, there exist positive integers n and m such that $[X] \approx [n]$ and $[Y] \approx [m]$. Using $k \cdot |Y| = |X|$, this means $km = n$. We define a k-to-1 correspondence $g : [n] \to [m]$ by splitting $[n] = [km]$ into an array with k columns and m rows as follows, and mapping the elements of each row i to $i \in [m]$:

$$\begin{array}{cccccccc}
1 & (m+1) & (2m+1) & \cdots & [(k-1)m+1] & \mapsto & 1 \\
2 & (m+2) & (2m+2) & \cdots & [(k-1)m+2] & \mapsto & 2 \\
3 & (m+3) & (2m+3) & \cdots & [(k-1)m+3] & \mapsto & 3 \\
\vdots & \vdots & \vdots & \ddots & \vdots & \vdots & \vdots \\
m & (m+k) & (2m+k) & \cdots & [(k-1)m+k] = km & \mapsto & m
\end{array}.$$

Using a bijection $h : X \to [n]$ and a bijection $k : [m] \to Y$, let $f : X \to Y$ be defined by $f = k \circ g \circ h$. It can be immediately verified that this composition is a k-to-1 correspondence. □

Problem 3.20. Show that the correspondence principle generalizes:

1. The bijection principle
2. The division principle

Example 3.21. Each vowel from the English language appears exactly once in the word SEQUOIA. How many permutations of this word are there such that, if the consonants are taken out of the permutation, the vowels appear in the same order EUOIA?

Solution. Without the restriction about the order of the vowels, there are 7! permutations. The vowels can be ordered in 5! ways if we do not consider consonants, so one in every 5! of the 7! permutations satisfies the restraint. Thus, the answer is

$$\frac{7!}{5!} = 7 \cdot 6 = 42.$$

□

Definition 3.22. If n and k are integers such that $1 \le k \le n$ and S is a set of n elements, then a **circular k-permutation** of S is an arrangement of k distinct elements of S in a circular formation. Unlike ordinary k-permutations, the positions are not indexed; instead, specifying a circular k-permutation is a matter of determining where each element lies relative to the others. More precisely, what distinguishes one circular k-permutation from another is:

- The underlying set of k elements
- For each element, what element is one position clockwise from it
- For each element, what element is one position counterclockwise from it

If it is necessary to differentiate circular k-permutations from the ordinary k-permutations, we will refer to the latter as **linear k-permutations**. Circular n-permutations are simply called **circular permutations**.

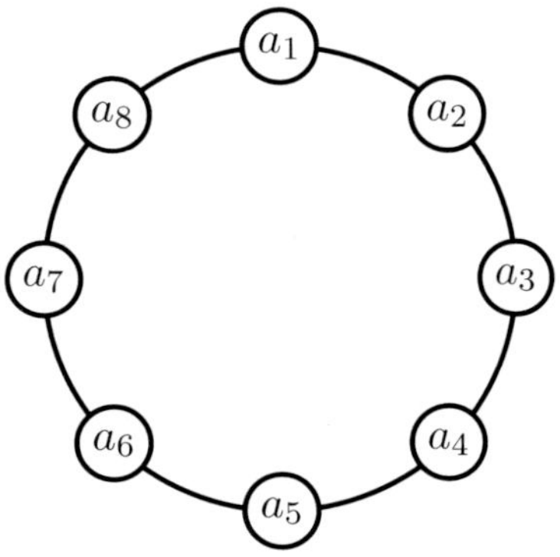

Theorem 3.23. If n and k are integers such that $1 \leq k \leq n$ and S is a set with n elements, then the set of circular k-permutations of S is finite with cardinality

$$\frac{P(n,k)}{k} = \frac{n!}{(n-k)! \cdot k}.$$

As a consequence, the set of circular permutations of S is finite with cardinality $(n-1)!$.

Proof. The idea is to identify each circular k-permutation with certain linear k-permutations. Specifically, we define a function that maps each linear k-permutation $(a_1, a_2, \ldots, a_k)$ of S to the circular permutation that goes clockwise from a_1 to a_2 and from a_2 to a_3, all the way through a_{k-1} to a_k, and finally from a_k to a_1. It is straightforward to see that each circular k-permutation has exactly k linear k-permutations mapping to it, making it a k-to-1 correspondence. By the correspondence principle, the set of circular k-permutations is finite with cardinality equal to the number of linear k-permutations divided by k, which is

$$\frac{P(n,k)}{k} = \frac{n!}{(n-k)! \cdot k}.$$

For $k = n$, this is $(n-1)!$. □

Problem 3.24. There are $n \geq 3$ keys that must be placed around a key ring. As with circular permutations, there is rotational symmetry, but a key ring can also be flipped around without changing the arrangement. How many such arrangements are there?

Circular permutations will be revisited in the context of counting necklaces using group theory in Section 12.2.

Definition 3.25. If n and k are integers such that $0 \leq k \leq n$, and S is a finite set of cardinality n, then a ***k*-combination** of S is a subset of S that has cardinality k. Equivalently, it is an element of $\mathcal{P}(S)$ of cardinality k.

Theorem 3.26. If n and k are integers such that $0 \leq k \leq n$, and S is a finite set of cardinality n, then the set of k-combinations of S is finite with cardinality $\dfrac{n!}{k!(n-k)!}$.

Proof. If $k = 0$ (or $n = 0$, in which case it is automatically true that $k = 0$ as well), then there is only one 0-combination of S, which is $\varnothing$. This matches the prescribed formula because

$$\frac{n!}{0!(n-0)!} = 1.$$

Now, we may assume that n and k satisfy $1 \leq k \leq n$. The idea is to map P_k, the set of k-permutations of S to C_k, the set of k-combinations of S, in a way that allows us to invoke the correspondence principle. Basically, we want to associate each k-permutation with the set of elements of S that underlie it by removing the order in the k-permutation but keeping the elements. Let's make this formal. Recall from Definition 3.6 that, for $k \geq 1$, a k-permutation of S is an injective function $p : [k] \to S$. So we define

$$\begin{aligned} f : P_k &\to C_k \\ p &\mapsto \operatorname{Rng}(p), \end{aligned}$$

where $\mathrm{Rng}(p)$ denotes the image or range of the function p. First we need to show that f actually maps each element of P_k to an element of C_k. The image of every function is surjective onto its range, and since p is injective to begin with, $p : [k] \to \mathrm{Rng}(p)$ is bijective. By the bijection principle, $\mathrm{Rng}(p)$ is indeed a k-element subset of S, and so it is an element of C_k. Thus, C_k can correctly be chosen to be a codomain of f.

Now, we want to prove that the correspondence principle applies to f. For each $c \in C_k$, f maps a permutation $p \in P_k$ to c if and only if $p : [k] \to S$ is an injective function with $\mathrm{Rng}(p) = c$. Since $|c| = k$, the injections $p : [k] \to c$ are precisely the permutations of c, the number of which we know to be $k!$. Thus, $f : P_k \to C_k$ is a $k!$-to-1 correspondence. Since P_k is known to be finite, the correspondence principle implies that C_k is finite with cardinality

$$|C_k| = \frac{|P_k|}{k!} = \frac{P(n,k)}{k!} = \frac{n!}{k!(n-k)!}.$$

□

Definition 3.27. For any integers k and n that satisfy $0 \le k \le n$, the number of k-combinations of a set of cardinality n is denoted by the notation

$$\binom{n}{k} = \frac{n!}{k!(n-k)!}.$$

The notation on the left side is called a **combination** or a **binomial coefficient**. The reason for the second name is due to a result called the binomial theorem, which we will introduce in Theorem 5.2.

As will become evident to the reader over time, combinations are even more common than permutations in combinatorics.

Example 3.28. One of several equivalent criteria for a polygon to be convex is that each line segment (excluding its endpoints) connecting each pair of non-adjacent vertices lies in the interior of the polygon. These line segments are called diagonals. Determine the number of diagonals of a convex polygon that has n edges.

Solution. If we remove the "non-adjacent" condition, then this larger set of "diagonals" is in bijection with the set of unordered pairs of two distinct vertices of the polygon. The latter set consists of the 2-element subsets of the set of vertices, of which there are $\binom{n}{2}$. Now, we have to subtract the number of unordered pairs of adjacent vertices. These correspond to the n edges of the polygon. Thus, the answer is

$$\binom{n}{2} - n = \frac{n(n-1)}{2} - n = \frac{n(n-3)}{2}.$$

A problem that the reader should ponder is: how many unordered pairs of these diagonals have an intersection point in the interior of the convex polygon? This is a more difficult problem that can also be solved by combinations. □

Problem 3.29. Let n and k be positive integers such that $n \ge 2k - 1$. Suppose there are k mutual enemies who are to be seated in n seats in such a way that no two people are sitting beside each other. In how many ways can this be done?

Problem 3.30. Given a rectangular grid measuring h units horizontally and v units vertically, determine the number of rectangles with all four edges landing on the grid lines.

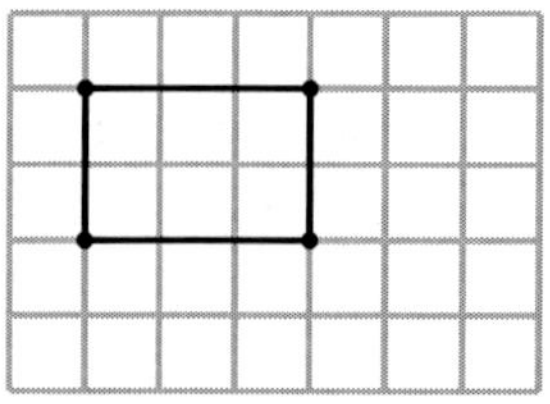

Problem 3.31. For integers p, q and a positive integer k such that $q - p + 1 \geq k$, let L be the set of lists $(c_1, c_2, \ldots, c_k)$ of k integers such that

$$p \leq c_1 < c_2 < \cdots < c_k \leq q.$$

Show that L is finite and determine its cardinality. This generalizes Problem 1.20.

A permutation of a multiset is defined formally as follows, but one would be better off thinking about it as a way of writing the elements of a multiset from left to right.

Definition 3.32. If $M : S \to \mathbb{Z}_+$ is a non-empty finite multiset, then a **permutation** of M is a list (see Definition 1.17) $\ell : [n] \to S$ such that:

1. n is the cardinality of M, which is the sum of the multiplicities of the elements in the support S of M.

2. For each $s \in S$, the number of entries of ℓ that are equal to s is the multiplicity of s in M. That is, $|\ell^{-1}(s)| = M(s)$.

Regarding the notation $\ell^{-1}(s)$, recall from Volume 1: If X and Y are non-empty sets and $f : X \to Y$ is a function, then the **preimage** under f of an element $y \in Y$ is the set

$$f^{-1}(y) = \{x \in X : f(x) = y\}.$$

Example. The lists $(1, 2, 3, 2)$ and $(2, 1, 3, 2)$ are permutations of the same multiset $\langle 1, 2, 2, 3 \rangle$. The reader may have noticed that, interestingly, we cannot write down a multiset in multiset notation without placing the elements in some order, even though order does not matter in a multiset. The same observation holds for sets in set notation. This is a limitation of written expression.

Theorem 3.33. Let M be a non-empty finite multiset, whose support is $\{a_1, a_2, \ldots, a_k\}$, and, for each index $1 \leq i \leq k$, the multiplicity of a_i is n_i. Then the set of permutations of M is finite with cardinality

$$\frac{(n_1 + n_2 + \cdots + n_k)!}{n_1! \cdot n_2! \cdots n_k!}.$$

Proof. The general idea is to move through the a_i and use combinations to choose indices at which to place a_i from among the indices that remain empty after the completion of previous steps. Formally, this utilizes the dependent multiplication principle.
Let n be the sum of the multiplicities, meaning

$$n = n_1 + n_2 + \cdots + n_k.$$

Let P be the set of permutations of M. In order to gather the ordering information contained in each permutation in an organized way, we map each permutation to a k-tuple of sets $(A_1, A_2, \ldots, A_k)$, where each A_i is the set of n_i indices at which a_i appears in the permutation. So the A_i are pairwise disjoint non-empty subsets of $[n] = \{1, 2, \ldots, n\}$ such that their union is $[n]$. Given such a k-tuple, we can uniquely construct a permutation of M that gets mapped to this k-tuple; it is simply a matter of placing each a_i in the indices assigned to it by the set A_i. Thus, the set of permutations of M is in bijection with the set of k-tuples $(A_1, A_2, \ldots, A_k)$ such that each A_i has the indices of the a_i's and the disjoint union of the A_i is $[n]$. Calling this set L, the bijection principle asserts that it suffices to show that L is finite with the prescribed cardinality.
In order to apply the dependent multiplication principle, we must show that L is symmetrically dependent and determine the dependence numbers. There are $\binom{n}{n_1}$ possible sets A_1, so this is the first dependence number. For each index $1 \leq i \leq k-1$ and for every valid assignment $(A_1, A_2, \ldots, A_i)$, there are

$$\binom{n - n_1 - n_2 - \cdots - n_i}{n_{i+1}} = \binom{n_{i+1} + n_{i+2} + \cdots + n_k}{n_{i+1}}$$

valid entries for A_{i+1} because $n_1 + n_2 + \cdots + n_i$ of the indices have been used up and we must choose n_{i+1} of the remaining indices in which to place n_{i+1} copies of a_{i+1}. This is the $(i+1)^{\text{th}}$ dependence number. Therefore, the dependence numbers exist, proving that the set is symmetrically dependent. By the dependent multiplication principle, L is finite with cardinality

$$\begin{aligned}|L| &= \binom{n_1 + \cdots + n_k}{n_1} \cdot \binom{n_2 + \cdots + n_k}{n_2} \cdots \binom{n_{k-1} + n_k}{n_{k-1}} \cdot \binom{n_k}{n_k} \\ &= \frac{(n_1 + \cdots + n_k)!}{n_1! \cdot (n_2 + \cdots + n_k)!} \cdot \frac{(n_2 + \cdots + n_k)!}{n_2! \cdot (n_3 + \cdots + n_k)!} \cdots \frac{(n_{k-1} + n_k)!}{n_{k-1}! \cdot n_k!} \cdot \frac{n_k!}{n_k! \cdot 0!} \\ &= \frac{(n_1 + n_2 + \cdots + n_k)!}{n_1! \cdot n_2! \cdots n_k!},\end{aligned}$$

where we used the formula for a binomial coefficient to go from the first line to the second, and telescoped the product in the second line to reach the third line. □

Definition 3.34. Let n be a non-negative integer, k be a positive integer, and $(n_1, n_2, \ldots, n_k)$ be a k-tuple of non-negative entries such that

$$n_1 + n_2 + \cdots + n_k = n.$$

Then the expression

$$\binom{n}{n_1, n_2, \ldots, n_k} = \frac{n!}{n_1! \cdot n_2! \cdots n_k!}$$

is called a **multinomial coefficient**. The reason for this mysterious name is due to the multinomial theorem, which is a result that we will discuss in Theorem 5.6.

Example. Multinomial coefficients are a generalization of binomial coefficients. Taking $k = 2$ shows that

$$\binom{n_1 + n_2}{n_1, n_2} = \frac{(n_1 + n_2)!}{n_1! n_2!} = \binom{n_1 + n_2}{n_1} = \binom{n_1 + n_2}{n_2},$$

where the left side is a multinomial coefficient and the right side expresses the left side as a binomial coefficient. In the other direction,

$$\binom{n}{k} = \frac{n!}{k!(n-k)!} = \binom{n}{k, n-k},$$

where the left side is a binomial coefficient and the right side expressed the left side as a multinomial coefficient.

As we discovered, when all of the n_i and $n = n_1 + n_2 + \cdots + n_k$ are positive, the multinomial coefficient above has a combinatorial interpretation as the number of permutations of the multiset that has support $\{a_1, a_2, \ldots, a_k\}$, where the multiplicity of each a_i is n_i. If some of the n_i are equal to 0, then the same interpretation does not hold because we do not allow elements to have a multiplicity of 0 in a multiset. However, this is merely a reflection of the inadequacy of the definition of a multiset for this purpose, and we could define an appropriate notion of a "generalized multiset" that allows for elements with zero multiplicity. We have not built a general theory around such a concept because there would be some nuances to consider, such as whether a generalized multiset counts as "finite" if its support is an infinite set but only finitely many elements of the support have non-zero multiplicity. However, we have commented on the case of a finite support as follows.

Definition 3.35. Let $\mathbb{Z}_{\geq 0}$ denote the set of non-negative integers.

Definition 3.36. A **multinomial set** is a function $f : S \to \mathbb{Z}_{\geq 0}$ such that its **support** S is a non-empty finite set. The image of each element of S is called the **multiplicity** of that element. The **cardinality** of a multinomial set is the sum of the multiplicities of the elements of its support. If the cardinality is n, then we call it an n**-multinomial set**. Two multinomial sets are said to be **equal** if each element that appears in a multinomial set a non-zero number of times appears the same number of times in the other multinomial set; interestingly, this means that two multinomial sets can be equal without having the same support since some elements could just have 0 multiplicity. A **permutation** of a multinomial set M of cardinality n is an n-tuple of the elements of M, where the number of times that each element of the support of M appears in the permutation is equal to the multiplicity of that element in M. In comparing this with the definition of a multiset (Definition 1.21), one should notice the difference between the codomains $\mathbb{Z}_+$ and $\mathbb{Z}_{\geq 0}$.

Example. If every element in the support of a multinomial set has multiplicity 0, then the only permutation of this multinomial set is the empty list ().

Theorem 3.37. Let M be a multinomial set whose support is $\{a_1, a_2, \ldots, a_k\}$ for some positive integer k, and for each index $1 \leq i \leq k$, the multiplicity of a_i is n_i. Analogous to the result for multisets, the set of permutations of M is finite with cardinality

$$\binom{n_1 + n_2 + \cdots + n_k}{n_1, n_2, \ldots, n_k} = \frac{(n_1 + n_2 + \cdots + n_k)!}{n_1! \cdot n_2! \cdots n_k!}.$$

Proof. We could attempt to deduce this from Theorem 3.33, but it would require showing that permutations of multinomial sets somehow correspond to permutations of certain multisets; they do so naturally but showing it is more trouble than it is worth. It is easier to replicate the argument used for permutations of multisets. Since we have already done it formally once in Theorem 3.33, we will describe it informally here. Let $n = n_1 + n_2 + \cdots + n_k$. If $n = 0$, then the fact that the only permutation is the empty list () matches the formula because 0! divided by a product of k copies of 0! is 1.
Now we many assume that $n \geq 1$, which is equivalent to saying that at least one of the n_i is positive. There are $\binom{n}{n_1}$ sets of n_1 indices at which the n_1 copies of a_1 could be placed, after which there are $\binom{n - n_1}{n_2}$ sets of n_2 indices at which the n_2 copies of a_2 could be placed, and so on until there is $\binom{n_k}{n_k} = 1$ set of n_k indices at which the n_k copies of a_k can be placed. By the dependent multiplication principle, we multiply these dependence numbers and telescope the product to get the answer, as before. □

Problem 3.38. Let a and b be non-negative integers and let $n = a + b$. Determine the number of n-tuples that consist of a entries equal to the symbol A and b entries equal to the symbol B. The reader should take note of this problem because it repeatedly appears in combinatorics.

Problem 3.39. Count the number of ways in which the composition

$$(g_n \circ g_{n-1} \circ \cdots g_2 \circ g_1) \circ f \circ (h_1 \circ h_2 \circ \cdots \circ h_{m-1} \circ h_m)$$

can be interpreted as a sequence of transformations starting with f. That is, we start from f and move outwards from it to the left towards g_n and to the right towards h_m in a way that each new function in the sequence is adjacent to a function that has previously appeared.

This brings us to the end of building basic combinatorial principles. We have proven them in an unusually formal fashion in the hope that the reader will have greater conviction about combinatorial truth than we did when learning combinatorics from other sources. In particular, instead of talking about "balls and boxes" from the physical universe, we have discussed constructions involving finite sets from the mathematical universe. Moreover, with the formalities over, the reader will not need to repeat the arguments; all that is needed is proof that a principle is applicable by showing that the hypotheses are satisfied. Of course, it is frequently not easy to decide which counting techniques should be applied and to determine whether the initial conditions hold in those techniques.

Chapter 4

Double Counting

"Not only is the above combinatorial proof much shorter than our previous proof, but it also makes the reason for the simple answer completely transparent. It is often the case, as occurred here, that the first proof to come to mind turns out to be laborious and inelegant, but that the final answer suggests a simpler combinatorial proof."

– *Richard Stanley, Enumerative Combinatorics I*

"“Invert, always invert.” It is in the nature of things, as Jacobi knew, that many hard problems are best solved only when they are addressed backward."

– *Charlie Munger, Harvard School Commencement Speech*

It can be pithily stated that, where bijective counting is about counting two sets in one way, double counting is about counting one set in two ways. We will see an example of this technique in relation to Pascal's triangle, and then explore proofs of a variety of standard combinatorial identities using this method. Our main two methods of producing double interpretations will be block-walking and forming committees.

4.1 Walking on Blocks

Ordinarily, we are happy to be able to find just one way of computing the cardinality of one of two equipotent finite sets. If we find two ways of counting a finite set, or if we are able to separately count two equipotent finite sets, then an interesting phenomenon occurs: we can set the two expressions equal to each other. If the two expressions are different and contain variables, then we have a **combinatorial identity**. We will not express this technique, called **double counting**, as a theorem because it is simply a novel application of the bijection principle, but this should not diminish its importance in the eyes of the reader.

Definition 4.1. If n is a positive integer, then a **lattice point** in n-dimensional Euclidean space $\mathbb{R}^n$ is a point such that all coordinates of the point are integers.

Definition 4.2. An **increasing path** on the Cartesian plane is a non-empty tuple of lattice points

$$((x_1, y_1), (x_2, y_2), \ldots, (x_n, y_n))$$

such that for each index $1 \leq i \leq n-1$, either $x_{i+1} = x_i + 1$ or $y_{i+1} = y_i + 1$, but not both. In geometric terms, we start at a point and move one unit right or one unit up at each step. For non-negative integers x and y, let $C(x, y)$ denote the set of increasing paths from $(0,0)$ to (x, y). Similarly, a **decreasing path** satisfies $x_{i+1} = x_i - 1$ or $y_{i+1} = y_i - 1$, but not both, for each index $1 \leq i \leq n-1$.

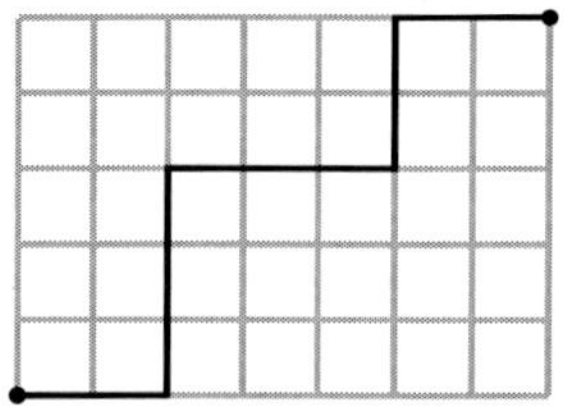

Example. A 1-tuple of coordinates $((x, y))$ is still a path. This path is both increasing and decreasing because there are no coordinates in the tuple other than the one pair. Also, note that traversing an increasing path in the opposite direction produces a decreasing path, and vice versa.

Definition 4.3. A **recursive relation** on a sequence of mathematical objects, such as numbers, is a function that expresses each object of sufficiently high index in terms of objects of lower indices.

Intuitively, recursion is about expressing the future solely in terms of what came in its past. We will discuss recursive sequences as a separate topic in Chapter 10 and Chapter 11.

Theorem 4.4 (Pascal's method)**.** For non-negative m and n, we can prove that $C(m,n)$ is finite and generate all of the cardinalities $|C(m,n)|$ as follows.

1. For all non-negative integers m and n, $C(m,0)$ and $C(0,n)$ are singletons. So
$$|C(m,0)| = |C(0,n)| = 1.$$
2. For all positive integers m and n, $C(m,n)$ is finite with cardinality
$$|C(m,n)| = |C(m-1,n)| + |C(m,n-1)|.$$

Thus, if we wish to place each number $|C(m,n)|$ at the coordinates (m,n), then we can iterate through the diagonals running from the top-left to the bottom-right and fill them in successively:

$$\begin{array}[b]{l} 1 \end{array} \qquad \begin{array}[b]{ll} 1 & \\ 1 & 1 \end{array} \qquad \begin{array}[b]{lll} 1 & & \\ 1 & 2 & \\ 1 & 1 & 1 \end{array} \qquad \begin{array}[b]{llll} 1 & & & \\ 1 & 3 & & \\ 1 & 2 & 3 & \\ 1 & 1 & 1 & 1 \end{array} \qquad \begin{array}[b]{lllll} 1 & & & & \\ 1 & 4 & & & \\ 1 & 3 & 6 & & \\ 1 & 2 & 3 & 4 & \\ 1 & 1 & 1 & 1 & 1 \end{array}$$

Proof. The first part will allow us to perform the recursion in the second part.

1. For any non-negative m, there is the path
$$((0,0),(1,0),\ldots,(m,0)).$$
Moreover, there can never be any steps up because it would not allow us to return to a y-coordinate of 0. So $|C(m,0)| = 1$. A symmetrical argument proves that
$$|C(0,n)| = 1.$$
2. Suppose m and n are positive integers. The set of increasing paths from $(0,0)$ to (m,n) can be partitioned into those with penultimate (second-last) vertex $(m-1,n)$ or $(m,n-1)$, as these are the only two lattice points that can lead to (m,n) in one step. The former set is in bijection with $C(m-1,n)$ and the latter set is in bijection with $C(m,n-1)$. By performing induction on the sum of the x-coordinate and y-coordinate, we can show that $C(m,n)$ is finite with cardinality
$$|C(m,n)| = |C(m-1,n)| + |C(m,n-1)|.$$

□

Pascal's method is applicable in more general scenarios, specifically directed graphs that do not contain any directed cycles. We will study graph theory in Chapter 8, but the standard proof of this statement is beyond our exposition, as it involves a concept called "topological ordering."

Theorem 4.5 (Block-walking)**.** Suppose x,y,x',y' are integers such that $x' \geq x$ and $y' \geq y$. Then the set of increasing paths that start at (x,y) and end at (x',y') is finite with cardinality
$$\frac{(x'-x+y'-y)!}{(x'-x)!\cdot(y'-y)!} = \binom{(x'-x)+(y'-y)}{(y'-y)}.$$

As a consequence, if m,n are non-negative integers then the number of increasing paths from $(0,0)$ to (m,n) is
$$\frac{(m+n)!}{m!\cdot n!} = \binom{m+n}{n}.$$

Thus, this is a geometric interpretation of all binomial coefficients.

Proof. It is not difficult to see that there must be exactly $x' - x$ steps right and $y' - y$ steps up, in some order. We can also see that every possible ordering of $x' - x$ steps right and $y' - y$ steps up yields a path from (x, y) to (x', y'). By setting up a bijection, it is equivalent to compute the number of ways of permuting $x' - x$ copies of the symbol R and $y' - y$ copies of the symbol U. Then Problem 3.38 yields the desired formula. The corollary about the number of increasing paths from $(0, 0)$ to (m, n) is the special case of $x = y = 0$ and $x' = m$ and $y' = n$. □

Corollary 4.6 (Pascal's identity)**.** For any positive integers n and k such that $n - 1 \geq k$,

$$\binom{n-1}{k-1} + \binom{n-1}{k} = \binom{n}{k}.$$

Proof. By combining our explicit block-walking formula (Theorem 4.5) with Pascal's method (Theorem 4.4), we get

$$\binom{n}{k} = |C(n-k, k)| = |C(n-k-1, k)| + |C(n-k, k-1)| = \binom{n-1}{k} + \binom{n-1}{k-1}.$$

This recursive relation between binomial coefficients has analogues for other combinatorial quantities. We will see them when we study distributions (see Problem 7.17, Problem 7.25, and Problem 7.32). □

Definition 4.7. By placing the number $|C(m, n)|$ at the point (m, n) on the Cartesian plane for all non-negative m and n, and rotating the plane by $135°$ clockwise, we get a triangular array where every binomial coefficient appears exactly once.

$$\begin{array}{ccccccccc}
 & & & & \binom{0}{0} & & & & \\
 & & & \binom{1}{0} & & \binom{1}{1} & & & \\
 & & \binom{2}{0} & & \binom{2}{1} & & \binom{2}{2} & & \\
 & \binom{3}{0} & & \binom{3}{1} & & \binom{3}{2} & & \binom{3}{3} & \\
\iddots & & \vdots & & \vdots & & \vdots & & \ddots
\end{array}$$

This is called **Pascal's triangle**. The row that has t as the upper entry of all of its binomial coefficients is called row t; for example, the top row is row 0. Sources often construct it by saying that the number at each position is to be found by summing the number to its top-left and the number to its top-right, which is equivalent to Pascal's method.

Example 4.8. Determine the number of increasing paths from $(0, 0)$ to (m, n) that do not pass through (p, q), where $0 \leq p \leq m$ and $\leq q \leq n$.

Solution. We will use complementary counting by counting the total number of increasing paths from $(0, 0)$ to (m, n) and then subtracting from it the number of increasing paths from $(0, 0)$ to (m, n) that *do* pass through (p, q). The total set is $C(m, n)$ which we know to have cardinality $\binom{m+n}{n}$. The smaller set is in bijection with the Cartesian product of the increasing paths from $(0, 0)$ to (p, q) and the increasing paths from (p, q) to (m, n). Thus, the unsuccessful paths number

$$\binom{p+q}{q} \cdot \binom{m-p+n-q}{n-q},$$

so the answer is

$$\binom{m+n}{n} - \binom{p+q}{q} \cdot \binom{m-p+n-q}{n-q}.$$

□

Problem 4.9. Let n be a positive integer, which is the number of steps in a process that we will describe. Objects A starts at $(0,0)$ and object B starts at (n,n) on the Cartesian plane. At each step of the process, A moves one unit up or one unit right but not both, and B moves one unit down or one unit left but not both. What is the probability that, after n steps, A and B occupy the same point?

Pascal's triangle contains numerous patterns. A simple one is that all of the triangular numbers

$$\binom{n}{2} = \frac{n(n-1)}{2} = 1 + 2 + \cdots + (n-1)$$

appear on the diagonal with binomial coefficients whose lower entries are equal to 2. Another pattern that we will now prove involves the Fibonacci numbers:

$$0, 1, 1, 2, 3, 5, 8, 13, 21, \ldots$$

which are recursively defined by $F_0 = 0$ and $F_1 = 1$ and $F_n = F_{n-1} + F_{n-2}$ for integers $n \geq 2$ (Definition 10.5).

Example 4.10. We left-justify Pascal's triangle to produce a right triangular array of numbers, and sum the diagonals from the bottom-left to the top-right as follows. One example of a diagonal has its entries and sum boxed.

$$\begin{array}{lllllll}
1 \\
1 & 1 \\
1 & 2 & 1 \\
1 & 3 & \boxed{3} & 1 \\
1 & \boxed{4} & 6 & 4 & 1 \\
\boxed{1} & 5 & 10 & 10 & 5 & 1 \\
1 & 6 & 15 & 20 & 15 & 6 & 1 \\
\vdots & & & & & & \ddots
\end{array}
\qquad
\begin{array}{rcl}
 & & 1 \\
 & & 1 \\
1+1 & = & 2 \\
1+2 & = & 3 \\
1+3+1 & = & 5 \\
1+4+3 & = & \boxed{8} \\
1+5+6+1 & = & 13 \\
 & \vdots &
\end{array}$$

This process seems to produce the Fibonacci numbers. Prove that it is indeed true that for integers $n \geq 1$,

$$F_n = \sum_{k=0}^{\lfloor \frac{n-1}{2} \rfloor} \binom{n-1-k}{k}.$$

Solution. We will prove the formula by double counting. We have been unable to motivate the set that will be double counted, but it is good to know in any case, as it is a well-known set. Let m be a positive integer. Recall from the solution to Example 3.2 that $\{0,1\}^m$ is the set of all binary strings of length m; in other words, they are all the m-tuples whose entries are only 0's and 1's. Let S_m be the subset of $\{0,1\}^m$ that contains no strings where two adjacent symbols are 1. We will double count S_m. Note that $\{0,1\}^m$ is finite with cardinality 2^m, so its subset S_m is also finite.

1. The first method is recursive. We begin by computing

$$S_1 = \{0,1\} \implies |S_1| = 2 = F_3,$$
$$S_2 = \{00, 01, 10\} \implies |S_2| = 3 = F_4.$$

For $m \geq 3$, we partition S_m into those strings whose rightmost symbol is 0 and those strings whose rightmost symbol is 1. The former set is in bijection with S_{m-1}, where removing the rightmost 0 is the bijection. The latter set is in bijection with S_{m-2} because the symbol directly to the left of the rightmost 1 must be 0, and so removing the 01 is the bijection. So $|S_1| = F_3$ and $|S_2| = F_4$ and we have just proven that

$$|S_m| = |S_{m-1}| + |S_{m-2}|$$

for all integers $m \geq 3$. Therefore, it holds by induction that $S_m = F_{m+2}$ for all integers $m \geq 1$.

2. In the context of the S_m, there is no meaning for F_1 or F_2, but it is easy to compute that

$$F_1 = 1 = \binom{0}{0},$$
$$F_2 = 1 = \binom{1}{0}.$$

It is also easy to verify that

$$F_3 = |S_1| = 2 = \binom{2}{0} + \binom{1}{1},$$
$$F_4 = |S_2| = 3 = \binom{3}{0} + \binom{2}{1}.$$

Now, we will count S_m for $m \geq 3$ using an ordinary counting argument, namely with binomial coefficients. For a given string in S_m, let i be the number of 0's and j be the number of 1's; this means $i + j = m$. There are $m + 1$ pairs (i, j) of non-negative integers such that $i + j = m$. However, not all of them lead to a non-empty set of strings in S_m. In particular, we claim that it is not possible that $i + 1 < j$. This is because if we lay out i copies of 0 in a row, then there are $i + 1$ positions in which we can distribute the j copies of 1: the $i - 1$ positions in between consecutive 0's, or directly to the left or the leftmost 0, or directly to the right of the rightmost 0. If $i + 1 < j$, then the pigeonhole principle asserts that some position will receive more than one copy of 1, which is not permissible. So, $i + 1 \geq j$. In all cases that $i + 1 \geq j$, there are $\binom{i+1}{j}$ ways to permute the i copies of 0 and j copies of 1 in a way that no two copies of 1 are adjacent, because we choose j of the described $i + 1$ slots.

Now note that since $i = m - j$, the condition $i + 1 \geq j$ is equivalent to $m - j + 1 \geq j$ or $\frac{m+1}{2} \geq j$. Since j is an integer, this is further equivalent to $\left\lfloor \frac{m+1}{2} \right\rfloor \geq j$, which explains the upper bound on the summation index. Therefore,

$$F_{m+2} = |S_m| = \sum_{j=0}^{\lfloor \frac{m+1}{2} \rfloor} \binom{i+1}{j} = \sum_{j=0}^{\lfloor \frac{m+1}{2} \rfloor} \binom{m+1-j}{j}$$

for $m \geq 3$. Combined with our computations of F_1, F_2, F_3, and F_4, this means that

$$F_n = \sum_{k=0}^{\lfloor \frac{n-1}{2} \rfloor} \binom{n-1-k}{k}$$

for all integers $n \geq 1$.

□

We will study the Fibonacci numbers closely in Section 10.2.

4.2 Forming Committees

Most of the identities in this section can be proven by mechanical algebraic or inductive methods, but a proof by double counting indicates a deeper understanding of the meaning of an identity. Double counting is about being given two formulas and reverse-engineering a family of finite sets such that the outputs of both formulas are their cardinalities. Instead of going from a structure to a formula, we are creating a structure that admits dual interpretations. In particular, there is a technique for doing this called committee formation, under which umbrella the standard proofs of several fundamental combinatorial identities fall. The idea is simple: given a set of people, assign special status to some of them.

Theorem 4.11. Let n and k be non-negative integers such that $n \geq k$. Then:

1. Pascal's triangle is symmetric across the central vertical line, meaning

$$\binom{n}{k} = \binom{n}{n-k}.$$

2. The sum of each row of Pascal's triangle can be computed as

$$\sum_{k=0}^{n} \binom{n}{k} = 2^n.$$

3. The sum of the squares of the elements of row n of Pascal's triangle is

$$\sum_{k=0}^{n} \binom{n}{k}^2 = \binom{2n}{n}.$$

Proof. Let $n \geq k$ be non-negative integers.

1. The set of k-subsets of n people is in bijection with the set of $(n-k)$-subsets of n people, where the bijection maps each way of choosing k people to the set of the $n-k$ people who have been excluded.

2. By the independent multiplication principle, 2^n is the number of ways of selecting a committee of any size (including the empty committee) out of n people (see Example 3.2). On the other hand, if the size of the committee is prescribed as k people, then there are $\binom{n}{k}$ possible committees. The possible committee sizes are $k = 0, 1, 2, \ldots, n$. The identity follows from summing these $n+1$ cases.

3. Sometimes, it is easier to apply a double counting argument after the desired identity has been algebraically transformed (in a reversible manner). By the symmetry of Pascal's triangle,

$$\sum_{k=0}^{n} \binom{n}{k}^2 = \sum_{k=0}^{n} \binom{n}{k}\binom{n}{n-k},$$

so it suffices to prove that

$$\sum_{k=0}^{n} \binom{n}{k}\binom{n}{n-k} = \binom{2n}{n}.$$

We accidentally proved this by block-walking in the solution to Problem 4.9, but there is an easier proof by committee-forming. Suppose there is a set of n distinct cats and n distinct dogs, and a committee of n animals must be selected from these $2n$ animals. In order to select n animals, there must be k cats and $n-k$ dogs for some $k \in \{0, 1, 2, \ldots, n\}$. For each such k, there are $\binom{n}{k}$ ways of choosing the cats and $\binom{n}{n-k}$ of choosing the dogs, and the two selections do not affect each other. So we add up all of the instances of $\binom{n}{k}\binom{n}{n-k}$, and set it equal to the direct count $\binom{2n}{n}$.

□

Theorem 4.12. For any positive integer n,

$$\sum_{\substack{0 \leq k \leq n \\ k \text{ even}}} \binom{n}{k} = \sum_{\substack{0 \leq k \leq n \\ k \text{ odd}}} \binom{n}{k} = 2^{n-1}.$$

A combinatorial interpretation of this is that the number of subsets of $[n]$ of even cardinality is equal to the number of subsets of $[n]$ of odd cardinality. Another way of writing the identity is

$$\sum_{k=0}^{n} (-1)^k \binom{n}{k} = 0.$$

Proof. This is saying that the sum of the alternating elements (even or odd indexed elements) in a row of Pascal's triangle is half the sum of that row. This is true because the sum of each row after row 0 is double the sum of the previous row, and by Pascal's identity, the sum of alternating elements of a row is the sum of the previous row. For example, in row 4,

$$\binom{4}{0} + \binom{4}{2} + \binom{4}{4} = \binom{3}{0} + \left[\binom{3}{1} + \binom{3}{2}\right] + \binom{3}{3} = 2^3 = 2^{4-1},$$
$$\binom{4}{1} + \binom{4}{3} = \left[\binom{3}{0} + \binom{3}{1}\right] + \left[\binom{3}{2} + \binom{3}{3}\right] = 2^3 = 2^{4-1}.$$

We have avoided providing a general argument because elements on the far left or far right equal to 1 require special consideration, and we would also have to do casework on the even and odd rows. The argument should be clear from the example above. We will provide an algebraic proof that does not require casework, using the binomial theorem in Theorem 5.20. □

Example 4.13. Prove that, if n, m, k are non-negative integers such that $n \geq m + k$, then

$$\binom{n}{m}\binom{n-m}{k} = \binom{n}{k}\binom{n-k}{m}.$$

Solution. Out of a set of n people, we will select an alpha committee of m people and a beta committee of k people such that the two committees are disjoint. If we select the people for the alpha committee first and then select the people for the beta committee, then the number of ways of doing so is $\binom{n}{m}\binom{n-m}{k}$. If we select the people for the beta committee first and then select the people for the alpha committee, then the number of ways of doing so is $\binom{n}{k}\binom{n-k}{m}$. These two computations count the same set, so the two expressions are equal. □

Problem 4.14. Prove the following identities:

1. If $n \geq k \geq m \geq 0$ are integers, then
$$\binom{n}{k} \cdot \binom{k}{m} = \binom{n}{m}\binom{n-m}{k-m}.$$

2. If $n \geq k \geq 1$ are integers, then
$$\binom{n}{k} = \frac{n}{k}\binom{n-1}{k-1}.$$

3. If $n \geq m \geq 0$ are integers, then
$$\sum_{k=m}^{n} \binom{n}{k}\binom{k}{m} = 2^{n-m}\binom{n}{m}.$$

Theorem 4.15 (Hockey stick identity). If n and k are non-negative integers such that $n \geq k$, then

$$\sum_{i=k}^{n} \binom{i}{k} = \binom{n+1}{k+1}.$$

Then name comes from the shape traced out by these binomial coefficients, including the one on the right side, in Pascal's triangle when $n > k$. Try it out!

Proof. There exists a proof by telescoping and a proof by induction, both of which utilize Pascal's identity, and we leave it to the reader to discover them. We will instead provide a proof by double counting. Selecting a committee of $k+1$ people out of $n+1$ people can be done in $\binom{n+1}{k+1}$ ways. There is an alternative method, which produces the sum on the left, that we will describe now. If $k = 0$, then it is easy to verify that

$$\sum_{i=k}^{n} \binom{i}{k} = \sum_{i=0}^{n} \binom{i}{0} = n + 1 = \binom{n+1}{1} = \binom{n+1}{k+1}.$$

Now we may assume that $n \geq k \geq 1$. First we place the $n + 1 \geq 2$ people in some order from left to right, such as ranking them by ascending height (this assumes without issue that no two of them have the same height). That is, we assign them unique indices from $[n + 1] = \{1, 2, \ldots, n, n + 1\}$. Then we do casework on the tallest person that appears in a committee of $k + 1 \geq 2$ people. The tallest person in the chosen committee cannot have index less than $k + 1$ because then there would be fewer than k shorter people available for the remaining k people on the committee. However, if the tallest person in the chosen committee has any of the indices $j = k + 1, k + 2, \ldots, n + 1$, then there are $j - 1 \geq k$ shorter people from which the k other people in the committee can be chosen. Adding up the cases, we get the identity

$$\binom{n+1}{k+1} = \sum_{j=k+1}^{n+1} \binom{j-1}{k} = \sum_{i=k}^{n} \binom{i}{k}.$$

□

Problem 4.16. It was asked in Volume 1 that a proof by telescoping be given for

$$1 \cdot 1! + 2 \cdot 2! + 3 \cdot 3! + \cdots + n \cdot n! = (n + 1)! - 1,$$

where n is any positive integer. Can you find a combinatorial proof by double counting? Simcha Waldman originally posed this question to the author.

Problem 4.17 (Reverse hockey stick identity). Use the ordinary hockey stick identity to prove that, if n and k are non-negative integers, then

$$\sum_{i=0}^{n} \binom{k+i}{i} = \binom{k+n+1}{n}.$$

The following is a relatively general identity that sometimes comes disguised as special cases of itself. This is because its three variables can be reduced to one or two variables, either by taking some of the variables to be constants or imposing a relation between some of the variables. We have shown a proof by double counting here, and algebraic proof is given in Theorem 5.9.

Theorem 4.18 (Vandermonde's identity). If m, n, k are non-negative integers such that $m + n \geq k$, then

$$\sum_{i=0}^{k} \binom{m}{i} \binom{n}{k-i} = \binom{m+n}{k},$$

where we use the convention that $\binom{p}{q} = 0$ if $p < q$.

Proof. Suppose there are m dogs and n cats, and we wish to select a committee of k animals from these $m + n$ dogs and cats. Then there can be $i = 0, 1, 2, \ldots, k$ dogs, and, in each such case, there must be $k - i$ cats. Adding up the cases yields

$$\sum_{i=0}^{k} \binom{m}{i} \binom{n}{k-i}.$$

Of course, the simpler method is to directly count the number of committees as $\binom{m+n}{k}$, which allows us to set the above sum equal to this expression. □

It may be asked what purpose combinatorial identities serve, outside of being interesting algebraic relations in their own right. In our experience, it happens frequently that the most direct and elegant way of computing the cardinality of a finite set does not necessarily come to mind immediately, and we must initially settle for a sub-optimal approach, such as casework. By using combinatorial identities, a messy expression can be simplified to produce an answer that is easier on the eye. For example, we need a combinatorial identity in the proof of the principle of inclusion-exclusion (Theorem 6.1). Moreover, sometimes the simplified expression holds insights into what a bijective proof might look like. This conviction has been expressed more eloquently by the combinatorialist, Richard Stanley, in his monumental two-volume work, *Enumerative Combinatorics.* The following is a problem that we independently developed where a combinatorial identity plays a simplifying role.

Example 4.19. Suppose n is a positive integer. A man simultaneously orders n books in a series, which are numbered $1, 2, 3, \ldots, n$. The n books arrive on n consecutive but separate days, with one book arriving per day. The sequence in which the books arrive is $k_1, k_2, k_3, \ldots, k_n$. He notices that every time a book k_i after k_1 arrives, there is some other book k_j that has already arrived such that k_i is the next book in the series after k_j or the preceding book in the series before k_j. More precisely, for each index $1 < i \leq n$, there exists an index $1 \leq j < i$ satisfying $|k_j - k_i| = 1$. In how many such ways can the books arrive?

Solution. We represent each way in which the books can arrive as an n-tuple $(k_1, k_2, \ldots, k_n)$. The first tactic is to partition the n-tuples according to which book arrives first. Given that $i \in [n] = \{1, 2, \ldots, n\}$ arrives first, there are $i - 1$ books with indices less than i that have yet to arrive and $n - i$ books with indices greater than i that have yet to arrive. Here is the key observation: in whichever order these $n - 1$ books arrive after the first book, the books $i + 1, i + 2, \ldots, n$ must arrive in ascending order and the books $i - 1, i - 2, \ldots, 1$ must arrive in descending order. Moreover, every sequence of arrivals that satisfy these two properties are suitable. So it is a matter of choosing $i - 1$ of the $n - 1$ spots to be the $i - 1$ books with indices less than i and they will fall into descending order; the remaining $n - i$ spots will be filled with the $n - i$ books with indices greater than i in ascending order. Thus, the answer is

$$\binom{n-1}{0} + \binom{n-1}{1} + \cdots + \binom{n-1}{n-1} = 2^{n-1}.$$

At the time of writing, we have been unable to find a direct counting argument for this problem, even though the final answer taunts us with its simple form. □

Problem 4.20. For each positive integer n, an **indecomposable permutation** of $[n]$ is defined as a permutation

$$(a_1, a_2, \ldots, a_n)$$

of $[n]$ such that, for all indices i such that $1 \leq i < n$, the first i entries $(a_1, a_2, \ldots, a_i)$ do not form a permutation of $[i]$. Denoting the number of indecomposable permutations of n by $f(n)$, use double counting to prove that

$$n! = \sum_{k=1}^{n} f(k)(n-k)!$$

holds for each positive integer n. Use this relation with $f(1) = 1$ to compute $f(7)$.

Double counting arguments have applications outside of purely combinatorial contexts. For example, in Volume 3, we will provide an exposition of the geometric double counting argument *par excellence* that Eisenstein formulated to prove the law of quadratic reciprocity.

Chapter 5

Algebraic Counting

"The bijective proofs give one a certain satisfying feeling that one 'really' understands why the theorem is true. The generating function arguments often give satisfying feelings of naturalness, and 'oh, I could have thought of that,' as well as usually offering the best route to finding exact or approximate formulas for the numbers in question."

– *Herbert S. Wilf, generatingfunctionology*

"Not everything that counts can be counted. Not everything that can be counted counts."

– *Albert Einstein*

Combinatorial identities are not always amenable to the method of double counting, in which case we can make use of algebraic methods. An algebraic structure that can be used to great effect in combinatorics is polynomials, which is a special cases of generating functions. It is also sometimes possible or necessary to prove identities by purely manipulative means that are devoid of any insight into the structures from which the identity might emerge.

5.1 Binomial and Multinomial Expansions

Problem 5.1. Expand $(x+y)^n$ and collect the like terms for $n = 0, 1, 2, 3, 4$. What do you notice about the coefficients?

Theorem 5.2 (Binomial theorem)**.** For any non-negative integer n,

$$(x+y)^n = \sum_{k=0}^{n} \binom{n}{k} x^k y^{n-k} = \sum_{k=0}^{n} \binom{n}{k} x^{n-k} y^k.$$

This explains why we call $\binom{n}{k}$ a "binomial coefficient" for non-negative integers n and k.

Proof. While it is possible to write an inductive proof, we will provide a more illuminating combinatorial proof. We are looking at the expansion of

$$(x+y)^n = \underbrace{(x+y)(x+y)\cdots(x+y)}_{n \text{ copies of } (x+y)}.$$

It can be proven by induction that the expansion, before like terms are collected, consists of the 2^n strings in the n-fold Cartesian product $\{x, y\}^n$. These are the n-tuples whose entries are all x's and y's. Upon each string being collapsed into the form $x^k y^{n-k}$, the possible values of k are $0, 1, 2, \ldots, n$. The coefficient of $x^k y^{n-k}$ is the number of permutations of k copies of x and $n-k$ copies of y, which is

$$\frac{(k+n-k)!}{k!(n-k)!} = \frac{n!}{k!(n-k)!} = \binom{n}{k}.$$

The binomial theorem follows from summing $\binom{n}{k} x^k y^{n-k}$ over $k = 0, 1, 2, \ldots, n$.

The second form with summands $\binom{n}{k}x^{n-k}y^k$ holds as well because we can iterate over the summands in the reverse order, and we can use the fact that $\binom{n}{n-k} = \binom{n}{k}$ to flip the binomial coefficients. □

Example 5.3. Inductively prove that, for each positive integer k, the sum of the kth powers of the first n positive integers

$$S_k(n) = 1^k + 2^k + 3^k + \cdots + n^k$$

can be expressed as a polynomial p_k with rational coefficients.

Solution. The following proof is due to Pascal. We will use the binomial theorem to extend a telescoping method that we used in Volume 1 to produce a formula for the sum of the first n squares. By telescoping, the binomial theorem, and the discrete Fubini's principle,

$$\begin{aligned}(n+1)^{k+1} - 1 &= \sum_{m=1}^{n} \left((m+1)^{k+1} - m^{k+1}\right) \\ &= \sum_{m=1}^{n}\sum_{t=0}^{k} \binom{k+1}{t} m^t \\ &= \sum_{t=0}^{k}\sum_{m=1}^{n} \binom{k+1}{t} m^t \\ &= \sum_{t=0}^{k} \binom{k+1}{t} (1^t + 2^t + \cdots + n^t).\end{aligned}$$

Thus, we can recursively determine $S_k(n)$ in terms of $S_t(n)$ for $0 \le t < n$, where we need the base case

$$S_0(n) = 1^0 + 2^0 + \cdots + n^0 = n.$$

This completes the proof, since the sum of polynomials is a polynomial. There is usually no need to explicitly find the formulas for $S_k(n)$ for $k \ge 4$ because they rarely crop up in practice. □

Problem 5.4. Note that

$$\left(x + \frac{1}{x}\right)^3 = x^3 + 3x + \frac{3}{x} + \frac{1}{x^3}.$$

We conclude that $x^3 + \frac{1}{x^3} = t^3 - 3t$, where $t = x + \frac{1}{x}$. Extend this idea by inductively proving that, for each positive integer n, $x^n + \frac{1}{x^n}$ can be expressed as $q_n(t)$ where q_n is a polynomial with integers coefficients.

Example 5.5. Suppose n is a positive integer. As a precursor to the multinomial theorem, expand

$$(x_1 + x_2 + \cdots + x_n)^2$$

and collect like terms.

Solution. Before like terms are collected, the expansion of

$$(x_1 + x_2 + \cdots + x_n)^2 = (x_1 + x_2 + \cdots + x_n)(x_1 + x_2 + \cdots + x_n)$$

consists of the n^2 terms $x_i x_j$ such that $(i, j) \in [n] \times [n]$. If $i = j$, then we get the n terms

$$x_1^2 + x_2^2 + \cdots + x_n^2.$$

If $i \neq j$, then $x_i x_j$ and $x_j x_i$ collect into the same term. Thus, the expression that we get after collecting like terms is

$$\sum_{k=1}^{n} x_k^2 + \sum_{i=1}^{n-1}\sum_{j=i+1}^{n} x_i x_j.$$

□

Theorem 5.6 (Multinomial theorem). For any positive integer m and any non-negative integer n,

$$(x_1 + x_2 + \cdots + x_m)^n = \sum_{\substack{k_1+k_2+\cdots+k_m=n \\ k_i \text{ non-negative}}} \binom{n}{k_1, k_2, \ldots, k_m} x_1^{k_1} x_2^{k_2} \cdots x_m^{k_m}.$$

This explains the etymology of the term "multinomial coefficient" in reference to expressions

$$\frac{(k_1 + k_2 + \cdots + k_m)!}{k_1! \cdot k_2! \cdots k_m!} = \binom{k_1 + k_2 + \cdots + k_m}{k_1, k_2, \ldots, k_m}$$

for non-negative k_i (Definition 3.34).

Proof. As with the binomial theorem, it is more illuminating to provide a combinatorial argument than an algebraic one. It can be shown by fixing a positive integer m and performing induction on positive integers n (the result is trivial for $n = 0$) that the expansion of $(x_1 + x_2 + \cdots + x_m)^n$, before like terms are collected, consists of the m^n elements of the n-fold Cartesian product $\{x_1, x_2, \ldots, x_m\}^n$. These are the n-tuples whose entries are chosen from $\{x_1, x_2, \ldots, x_m\}$, possibly with repetition. Note that each string can be collapsed into the form $x_1^{k_1} x_2^{k_2} \cdots x_m^{k_m}$ where the exponents k_i are non-negative integers that sum to n. The coefficient of this term is the number of permutations formed by k_1 copies of x_1, k_2 copies of x_2, and so on, up to k_m copies of x_m. By the definition of a multinomial coefficient, this number is

$$\frac{(k_1 + k_2 + \cdots + k_m)!}{k_1! \cdot k_2! \cdots k_m!} = \binom{n}{k_1, k_2, \ldots, k_m}.$$

The multinomial theorem is produced by adding up the terms $\binom{n}{k_1, k_2, \ldots, k_m} x_1^{k_1} x_2^{k_2} \cdots x_m^{k_m}$ over all m-tuples $(k_1, k_2, \ldots, k_m)$ of non-negative integers that sum to n. □

In the binomial expansion of $(x + y)^n$, it is clear that there are $n + 1$ terms after like terms are collected. It is not as obvious how many terms are in a multinomial expansion after like terms are collected. We will revisit this problem using the "sticks and stones" method for distributions in Problem 7.12.

5.2 Generating Functions

Recall from Volume 1 that a polynomial can be interpreted in two ways. One is as a formal polynomial, where the variables are indeterminates (i.e. symbols that have no value and exist for organizational purposes). The other is as a function where values can be substituted in for the variables. We will look at applications of both perspectives in combinatorics and their interchangeability. Two formal polynomials are said to be equal if and only if the coefficients of corresponding terms are equal. This technique of comparing coefficients of equal formal polynomials appears from time to time in discrete mathematics, such as in a proof of Lucas's congruence for binomial coefficients in number theory and in a weak variant of Wolstenholme's theorem, both provided in Volume 3; we will see it momentarily in an algebraic proof of Vandermonde's identity (Theorem 5.9). Let us see what exactly it is that we are doing when we evaluate a polynomial in two different ways and equate corresponding coefficients. It is essentially a matter of the associativity of the multiplication of generating functions.

Definition 5.7. A **generating function**, is a countable sequence of complex coefficients $(a_0, a_1, a_2, \ldots)$, indexed by the non-negative integers and presented as a formal sum

$$\sum_{k=0}^{\infty} a_k x^k = a_0 + a_1 x + a_2 x^2 + a_3 x^3 + \cdots.$$

By formal, we mean that the symbol x is an indeterminate that is used for organizing a sequence and does not necessarily represent a complex number, though it is certainly possible (and often desirable) to substitute in a number in cases where it leads to a convergent sum. Two generating functions are said to be **equal** if they are equal as sequences, meaning entries at the same indices are equal. If all terms of sufficiently large index are 0, then we have a **polynomial**. If $(a_i)_{i\geq 0}$ and $(b_i)_{i\geq 0}$ are generating functions, then their **sum** is a generating function that is produced by component-wise addition $(a_i+b_i)_{i\geq 0}$. Following the distributive law, their **product** is defined as $(c_i)_{i\geq 0}$ where

$$c_i = a_0b_i + a_1b_{i-1} + \cdots + a_{i-1}b_1 + a_ib_0$$

for all non-negative integers i. We define positive integer **powers** of generating functions according to repeated multiplication, and the 0th power of any non-zero generating function to be the constant 1 (all higher entries are 0).

Theorem 5.8. The multiplication of generating functions is associative, meaning if A, B, C are generating functions then

$$A \cdot (B \cdot C) = (A \cdot B) \cdot C.$$

As a consequence, regrouping allows us to say that, for all non-negative integers m, n,

$$\begin{aligned} A^{m+n} &= A^m \cdot A^n, \\ A^{mn} &= (A^m)^n. \end{aligned}$$

Thanks to the commutativity of the multiplication of generating function, which is an evident property, we also have

$$(AB)^n = A^n \cdot B^n.$$

These are tools for expanding polynomials or generating functions in two ways and then equating their coefficients, thereby possibly deriving a non-trivial algebraic identity.

Proof. To prove the associativity of the multiplication of generating functions, we must follow the definition of "multiplication" here carefully, as it is a formal definition and we are not dealing with ordinary arithmetic, even though the former is inspired by the latter. Our proof will be highly reminiscent of the proof of associativity of the Dirichlet convolution from number theory in Volume 3. Let

$$A = (a_i)_{i\geq 0}, B = (b_i)_{i\geq 0}, C = (c_i)_{i\geq 0}$$

be three generating functions. For each generating function G, we will denote the coefficient of each x^t by G_t. Below, all sums of the form $\sum\limits_{u+v=w}$ are done by iterating over all non-negative pairs (u, v) that sum to w, and similarly for triples. Then

$$(A \cdot (B \cdot C))_t = \sum_{i+\ell=t} a_i(B \cdot C)_\ell = \sum_{i+\ell=t} a_i \sum_{j+k=\ell} b_jc_k = \sum_{i+\ell=t}\sum_{j+k=\ell} a_i(b_jc_k).$$

So we have a double sum that first splits t into $i+\ell$ and then splits ℓ into $j+k$. This is the same as splitting t into $i+j+k$ from the beginning and cutting out the middleman ℓ. Using the associativity of complex numbers, the sum is equal to

$$\sum_{i+j+k=t} a_i(b_jc_k) = \sum_{i+j+k=t} (a_ib_j)c_k.$$

Similarly, the expansion of the other side of the associativity identity is

$$\begin{aligned} ((A \cdot B) \cdot C)_t = \sum_{\ell+k=t} (A \cdot B)_\ell c_k &= \sum_{\ell+k=t} \left[\left(\sum_{i+j=\ell} a_ib_j\right) c_k\right] \\ &= \sum_{\ell+k=t}\sum_{i+j=\ell} (a_ib_j)c_k = \sum_{i+j+k=t} (a_ib_j)c_k. \end{aligned}$$

Since this is true for every non-negative integer t, it holds that

$$A \cdot (B \cdot C) = (A \cdot B) \cdot C.$$

The subsequent identities about A^{m+n} and A^{mn} and $(AB)^n$ are easy enough to see intuitively: associativity allows us to apply the multiplication operation in any order, and these identities are about regrouping the generating functions in the product in various ways. ☐

Theorem 5.9 (Vandermonde's identity)**.** If m, n, k are non-negative integers such that $m + n \geq k$, then it can be proven that

$$\sum_{i=0}^{k} \binom{m}{i}\binom{n}{k-i} = \binom{m+n}{k}$$

by expanding

$$(x+1)^{m+n} = (x+1)^m(x+1)^n$$

in two different ways. A double counting proof was given in Theorem 4.18.

Proof. The two expansions are

$$\sum_{t=0}^{m+n} \binom{m+n}{t} x^t = (x+1)^{m+n} = (x+1)^m(x+1)^n = \left[\sum_{i=0}^{m} \binom{m}{i} x^i\right] \cdot \left[\sum_{j=0}^{n} \binom{n}{j} x^j\right].$$

Upon expansion but before like terms are collected, the product at the end has terms of the form

$$\binom{m}{i}\binom{n}{j} x^{i+j}.$$

We will group them according to fixed values of $t = i+j$. The possible values of $t = i+j$ are $0, 1, 2, \dots, m+n$. For a fixed t, the possible ordered pairs (i, j) are $(0, t), (1, t-1), (2, t-2), \dots, (t, 0)$. So, we would like to say that we can collect the like terms and get

$$\sum_{t=0}^{m+n} \left(\sum_{\ell=0}^{t} \binom{m}{\ell}\binom{n}{t-\ell}\right) x^t.$$

At first, this might seem to have an excess of terms because the factor $\sum_{i=0}^{m} \binom{m}{i} x^i$ does not contain any terms with $i > m$ and the factor $\sum_{j=0}^{n} \binom{n}{j} x^j$ does not contain any terms with $j > n$, whereas we cannot always make the claim that the coefficients $\binom{m}{\ell}\binom{n}{t-\ell}$ have $\ell \leq m$ and $t - \ell \leq n$ because t iterates from 0 through $m+n$. However, this is not a problem because the convention that $\binom{p}{q} = 0$ for $p < q$ annihilates those terms.

For each $k \in [m+n]*$, equating the coefficient of x^k in both expansions yields

$$\binom{m+n}{k} = \sum_{\ell=0}^{k} \binom{m}{\ell}\binom{n}{k-\ell},$$

which is the identity that we seek. □

Generating functions can be utilized in a variety of combinatorial problems ranging from partitions to recurrences. Rigorously studying generating functions requires the tools of calculus, due to the issue of convergence and needing to take derivatives, among other reasons. Let us begin with generating functions where all terms of sufficiently large degree have coefficients equal to 0, making them equivalent to formal polynomials. In particular, we will show that formal polynomials, and generating functions by extension, have a peculiar affinity for solving certain combinatorial problems that are difficult to approach by any of the other methods explored thus far.

Example 5.10. Alice has coins worth $1, 2, 3, 4$ dollars, Bob has coins worth $2, 3, 5$ dollars and Cathy has coins worth $2, 3, 5, 8$ dollars. In how many ways can 6 dollars be produced if exactly one coin is collected from each person?

Solution. If a dollars is collected from Alice, b dollars is collected from Bob, and c dollars is collected from Cathy, then we represent this way of collecting money by the term $x^a x^b x^c$, where x is an indeterminate. The key idea is to use the generating function

$$(x^1 + x^2 + x^3 + x^4)(x^2 + x^3 + x^5)(x^2 + x^3 + x^5 + x^8),$$

where the reasoning behind the choice of the exponents of each factor comes from the denominations that the corresponding person has available. After expansion, but before like terms are collected, the terms are precisely the set of $x^a x^b x^c$ such that (a, b, c) is an element of the Cartesian product

$$\{1, 2, 3, 4\} \times \{2, 3, 5\} \times \{2, 3, 5, 8\}.$$

We are seeking the number of triples in this Cartesian product such that $a+b+c = 6$. Thus, it is equivalent to find the coefficient of x^6 in the expansion after each term $x^a x^b x^c$ is written as x^{a+b+c} and like terms are collected. At this point, we could proceed with expansion, but we can avoid some of the messy work by making the observation that the polynomial factors as

$$x^5(1 + x + x^2 + x^3)(1 + x + x^3)(1 + x + x^3 + x^6).$$

Thanks to the x^5 factor, we are seeking the coefficient of x in the expansion of

$$(1 + x + x^2 + x^3)(1 + x + x^3)(1 + x + x^3 + x^6).$$

We do not need to expand it completely because the only ways of producing 1 by adding three ordered non-negative integers (which are all possible in this case) are

$$1 + 0 + 0 = 0 + 1 + 0 = 0 + 0 + 1 = 1.$$

Therefore, the answer is 3. If we are feeling nervous, we can construct the solutions

$$2 + 2 + 2 = 1 + 3 + 2 = 1 + 2 + 3 = 6,$$

but we will still have to trust the proof that there are no other solutions lurking. In any case, generating functions are not devised for constructing solutions. We were able to construct the solutions in this case only because the problem is a basic example. □

Problem 5.11. In a box lies one red ball, one orange ball, two yellow balls, and three green balls. Balls of the same colour are indistinguishable and balls of different colours are distinguishable. In how many ways can three (unordered) balls be collected from the box?

Problem 5.12. A tetrahedral die is a regular tetrahedron that has the numbers $1, 2, 3, 4$ indicated on the four faces, with one number per face. When a tetrahedral die is thrown, the number that is "rolled" is the number on the hidden face that is parallel to the ground. Three tetrahedral dice, coloured red, white, and blue, so that they are distinguishable, are thrown. In how many ways can the sum of the three rolled numbers be 4 or 5?

Problem 5.13. For any non-negative integer t and positive integer b, prove the combinatorial identity

$$\sum_{i=0}^{t} (-1)^i \binom{t}{i} \binom{bt - bi}{t} = b^t.$$

Before moving on from generating functions, let us observe their true strength in the area of infinite power series. There are two critical ideas:

1. If we switch to interpreting the indeterminate x as a variable that takes on values in some interval around 0, then the power series (which now becomes a infinite series of real numbers) might converge in that interval. This could allow for a succinct expression of the series. By and large, we will only deal with infinite geometric series and their variations since those are the only infinite series that we are comfortable with summing (see Volume 1).

2. If two power series converge on the same non-empty interval around 0, and the same inputs lead to the same convergent outputs for the two series on that interval, then the coefficients of corresponding terms are equal. This powerful idea (pun unintended) may be proven by taking successive derivatives of the power series and substituting in 0.

The interesting concept that we will now explore, in particular in the area of partitions, is that two seemingly different generating functions, each obtained in a disguised manner through the product of other generating functions (as in the finite cases above), converge to the same function on some interval around 0, and therefore they must have the same coefficients!

Definition 5.14. A **partition** of a positive integer n is a multiset of positive integers whose sum is n. So, order does not matter in this collection and repeated elements are allowed. Each element of the multiset is called a **part**; the **number of parts** in a partition is the sum of the multiplicities of all elements of the support, not the number of elements of the support. For example, $\langle 1, 2, 2, 2, 3, 3 \rangle$ has 5 parts.

Theorem 5.15. *For each positive integer n, let $p(n)$ denote the number of partitions of n. Then the generating function for p may be written as*

$$\sum_{n=0}^{\infty} p(n)x^n = \prod_{k=1}^{\infty} (1 + x^k + x^{2k} + x^{3k} + \cdots) = \prod_{k=1}^{\infty} \frac{1}{1 - x^k},$$

where we temporarily define that $p(0) = 1$ to make it all work out nicely. The product in the middle does in fact expand out to a generating function.

Proof. The product

$$\prod_{k=1}^{\infty} (1 + x^k + x^{2k} + x^{3k} + \cdots) = \lim_{t \to \infty} \prod_{k=1}^{t} (1 + x^k + x^{2k} + x^{3k} + \cdots)$$

actually equals a generating function $\sum_{n=0}^{\infty} f(n)x^n$ for some function f on $\mathbb{Z}_{\geq 0}$ because after multiplying out the first t multiplicands, multiplying the result by any higher multiplicands does not alter the summands $f(n)x^n$ up to and including $f(t)x^t$. So each additional multiplicand that we bring into play guarantees that an additional summand $f(n)x^n$ is fixed. In this way, the infinite product (which is a limit of finite products) produces a generating function. Moreover, by the formula for a geometric series,

$$\prod_{k=1}^{\infty} (1 + x^k + x^{2k} + x^{3k} + \cdots) = \prod_{k=1}^{\infty} \frac{1}{1 - x^k}.$$

Now we must show that the generating function produced is the one for counting partitions. This is a matter of noticing that each multiplicand

$$1 + x^k + x^{2k} + x^{3k} + \cdots$$

provides choices about how many times k appears in a partition. In the expansion, the coefficient of x^n will contain the number of distinct ways of obtaining n as a sum of so-and-so many copies of various positive integers; every possible partition of n is accounted for in this way. □

This establishes in our minds the sort of correspondence that is typically used to solve partitions problems using generating functions. Admittedly, we have not proven that this series converges in some interval around 0. We will soon commit more such forgivable misdeeds by telescoping infinite products. Students who take courses in rigorous calculus will be able to fill in the gaps. The following is a classic result to kick off our study of identities involving partitions.

Theorem 5.16 (Euler's partitions theorem)**.** *For each positive integer n, the number of partitions of n into distinct parts is equal to number of partitions of n into parts that are each odd.*

Proof. The generating function for the number of partitions into distinct parts is $\prod_{k=1}^{\infty}(1+x^k)$ and the generating function for the number of partitions into parts that are each odd is

$$\prod_{k=0}^{\infty}(1+x^{2k+1}+x^{2(2k+1)}+x^{3(2k+1)}+\cdots).$$

We want to prove that these are equal. Miraculously, we can use the difference of squares factorization and telescoping to say that

$$\prod_{k=1}^{\infty}(1+x^k)=\frac{\prod_{k=1}^{\infty}(1-x^{2k})}{\prod_{k=1}^{\infty}(1-x^k)}=\prod_{k=0}^{\infty}\frac{1}{1-x^{2k+1}},$$

where in the last step we were able to cancel out every factor in the numerator with alternating factors in the denominator. By the formula for an infinite geometric series, each multiplicand on the right side is equal to

$$\frac{1}{1-x^{2k+1}}=1+x^{2k+1}+x^{2(2k+1)}+x^{3(2k+1)}+\cdots,$$

which matches the multiplicands of the second generating function. Thus, the generating functions are equal as functions, which allows us to say that the two sequences enumerated are equal. □

Example 5.17. For each positive integer n, prove that:

1. The number of partitions of n into parts that are all congruent to $\pm 1 \pmod 6$ is equal to the number of partitions of n into distinct parts that are all congruent to $\pm 1 \pmod 3$.
2. The number of partitions of n such that no part appears exactly once is equal to the number of partitions of n where each part is divisible by 2 or 3 or both.

Solution. We will prove the first part and then use it as a lemma for the second part.

1. Again, by the difference of squares factorization and the formula for a geometric series,

$$\begin{aligned}\prod_{k=0}^{\infty}(1+x^{3k+1})(1+x^{3k+2})&=\frac{\prod_{k=0}^{\infty}(1-x^{6k+2})(1-x^{6k+4})}{\prod_{k=0}^{\infty}(1-x^{3k+1})(1-x^{3k+2})}\\&=\frac{\prod_{k=0}^{\infty}(1-x^{6k+2})(1-x^{6k+4})}{\prod_{k=0}^{\infty}(1-x^{6k+1})(1-x^{6k+4})(1-x^{6k+2})(1-x^{6k+5})}\\&=\frac{1}{\prod_{k=0}^{\infty}(1-x^{6k+1})(1-x^{6k+5})}.\end{aligned}$$

   The reader should check that the generating functions on the far left and far right are in fact the two that we want to equate.

2. Being divisible by 2 or 3 means not being congruent to $\pm 1 \pmod 6$. Moreover, note that

$$1+(x^{2k}+x^{3k}+x^{4k}+\cdots)=1+\frac{x^{2k}}{1-x^k}.$$

   So, we are seeking to establish the equality

$$\prod_{k=1}^{\infty}\left(1+\frac{x^{2k}}{1-x^k}\right)=\left[\frac{\prod_{k=1}^{\infty}(1-x^k)}{\prod_{k=0}^{\infty}(1-x^{6k+1})(1-x^{6k+5})}\right]^{-1}=\frac{\prod_{k=0}^{\infty}(1-x^{6k+1})(1-x^{6k+5})}{\prod_{k=1}^{\infty}(1-x^k)}.$$

   By taking the reversible step of multiplying both sides by $\prod_{k=1}^{\infty}(1-x^k)$, it is equivalent to prove that

$$\prod_{k=1}^{\infty}(1-x^k+x^{2k})=\prod_{k=0}^{\infty}(1-x^{6k+1})(1-x^{6k+5}).$$

By the reciprocal of the previous part of the problem and the sum of cubes factorization,

$$\begin{aligned}\prod_{k=0}^{\infty}(1-x^{6k+1})(1-x^{6k+5}) &= \frac{1}{\prod_{k=0}^{\infty}(1+x^{3k+1})(1+x^{3k+2})} \\ &= \frac{\prod_{k=1}^{\infty}(1+x^{3k})}{\prod_{k=1}^{\infty}(1+x^{k})} = \prod_{k=1}^{\infty}\frac{1+x^{3k}}{1+x^{k}} \\ &= \prod_{k=1}^{\infty}(1-x^{k}+x^{2k}),\end{aligned}$$

which proves the desired identity.

□

Problem 5.18. For each positive integer n, prove that the number of partitions of n into parts that are non-multiples of 3 is equal to the number of partitions of n into parts where each part appears $0, 1,$ or 2 times and no more. See if the method used can lead to a more general result.

Problem 5.19. For each positive integer n, prove that the number of partitions of n in which each part m appears strictly fewer than m times is equal to the number of partitions of n in which no part is a square.

There is an endless supply of such curious patterns among partitions that may be proven by generating functions, but what we have shown captures their general essence. If it seems that our excursion into partitions was short-lived, the reader will be interested in knowing that partitions will reappear in Section 7.3. As a final note, we want to make it clear that these identities are equalities between cardinalities of sets and not the sets themselves.

5.3 Direct Computation

Thus far, we have largely focused on methods of proving combinatorial identities that provide some degree of insight into what mathematical structures cause these identities to materialize. However, this is not always feasible. As a way of easing into a direct computation approach, we will switch from formal polynomials to interpreting polynomials as functions where the variables are replaced by real or complex values and the expression is evaluated according to the rules of arithmetic.

Theorem 5.20. The following sums hold:

1. If n is a non-negative integer, then $\sum_{k=0}^{n}\binom{n}{k} = 2^n$.

2. If n is a non-negative integer, then $\sum_{k=0}^{n}(-1)^k\binom{n}{k} = 0$.

3. If n is a positive integer, then

$$\sum_{k=0}^{\infty}\binom{n}{2k} = \sum_{k=0}^{\infty}\binom{n}{2k+1} = 2^{n-1}.$$

These "infinite" sums make sense because all terms of sufficiently high index are 0. The notation

$$\sum_{k=0}^{\infty}a_k = a_0 + a_1 + a_2 + \cdots$$

indicates that we are too lazy to find the last meaningful index of the series, or perhaps it is too inconvenient to do so.

Proof. We will make substitutions into the binomial theorem, which states that

$$(x+y)^n = \sum_{k=0}^{n} \binom{n}{k} x^k y^{n-k}.$$

1. For $x = y = 1$, we get

$$2^n = (1+1)^n = \sum_{k=0}^{n} \binom{n}{k}.$$

   The corresponding theorem that results from substituting 1's in for all of the variables in the multinomial theorem will make its appearance in Cayley's formula for labelled trees in Problem 8.13.

2. For $x = -1$ and $y = 1$, we get

$$0 = (-1+1)^n = \sum_{k=0}^{n} (-1)^k \binom{n}{k}.$$

3. By the second part,

$$0 = \sum_{k=0}^{n} (-1)^k \binom{n}{k} = \sum_{k=0}^{\infty} \binom{n}{2k} - \sum_{k=0}^{\infty} \binom{n}{2k+1},$$

   so $\sum_{k=0}^{\infty} \binom{n}{2k} = \sum_{k=0}^{\infty} \binom{n}{2k+1}$. By the first part, the left and right sides add up to 2^n, so these equal sums are each $\frac{2^n}{2} = 2^{n-1}$.

□

Problem 5.21. For all integers $n \geq 5$, prove by induction that $\binom{2n}{n} < 4^{n-1}$.

Problem 5.22. For all positive integers n, prove that

$$\binom{n}{0} + \binom{n}{4} + \binom{n}{8} + \cdots = 2^{n-2} + 2^{\frac{n-2}{2}} \cdot \cos \frac{n\pi}{4}.$$

The following hints should help:

1. The fourth roots of unity are $1, i, -1, -i$. Substitute them for x into $(1+x)^n$, stack the four binomial expansions on top of each other and cancel out terms that are negatives of each other. What is left?

2. Evaluating $(1+x)^n$ for $x = 1$ and $x = -1$ is easy (see Theorem 5.20). For $x = i$ and $x = -i$, rewrite $(1+x)^n$ using de Moivre's formula from Volume 1.

3. Finally, equate the two expressions for $(1+1)^n + (1+i)^n + (1-1)^n + (1-i)^n$. Simplify as needed.

Now we will show how Problem 5.22 can be generalized.

Theorem 5.23. Suppose m and n are positive integers. Then

$$\sum_{k=0}^{\infty} \binom{n}{km} = \frac{2^n}{m} \cdot \sum_{k=1}^{m} \left[\left(\cos \frac{k\pi}{m} \right)^n \cdot \cos \frac{nk\pi}{m} \right].$$

Proof. Let m and n be positive integers. Let

$$\zeta_k = e^{\frac{2k\pi}{m} i} \text{ for } k = 1, 2, \ldots, m.$$

These are the mth roots of unity. By the binomial theorem,

$$\sum_{k=1}^{m}(1+\zeta_k)^n = \sum_{k=1}^{m}\sum_{j=0}^{n}\binom{n}{j}\zeta_k^j = \sum_{j=0}^{n}\binom{n}{j}(\zeta_1^j+\zeta_2^j+\cdots+\zeta_m^j),$$

Note that we have used the discrete Fubini's principle to switch the order in which the summations occurred. Now we will evaluate each summand, according to whether m divides j or m does not divide j.
If m divides j, then let p be the integer $p=\dfrac{j}{m}$. Then

$$\zeta_1^j+\zeta_2^j+\cdots+\zeta_m^j = \sum_{k=1}^{m}\zeta_k^{mp} = \sum_{k=1}^{m}(\zeta_k^m)^p = \sum_{k=1}^{m}1^p = m.$$

If m does not divide j, then we seek to find the sum

$$\sum_{k=1}^{m}\zeta_k^j = \sum_{k=1}^{m}(\zeta_1^k)^j = \sum_{k=1}^{m}(\zeta_1^j)^k,$$

which is a geometric series. The formula for a geometric series is

$$1+z+z^2+\cdots+z^t = \frac{z^{t+1}-1}{z-1}$$

for any non-negative integer t, as long as z is a complex number that is not equal to 1. Suppose for contradiction that $\zeta_1^j=1$. This means $e^{\frac{2j\pi}{m}i}=1$, which implies that $\dfrac{2j\pi}{m}\equiv 0 \pmod{2\pi}$. So there exists an integer q such that $\dfrac{2j\pi}{m}=2\pi q$. Simplifying, we get $j=qm$, which is a contradiction because we are working on the case that m does not divide j. So $\zeta_1^j\neq 1$. Thus, we can apply the formula for a geometric series to get

$$\sum_{k=1}^{m}(\zeta_1^j)^k = \frac{(\zeta_1^j)^{m+1}-1}{\zeta_1^j-1}-1 = \frac{(\zeta_1^j)^{m+1}-\zeta_1^j}{\zeta_1^j-1} = \zeta_1^j\cdot\frac{(\zeta_1^j)^m-1}{\zeta_1^j-1} = 0,$$

where the numerator disappears in the end due to the fact that

$$(\zeta_1^j)^m-(\zeta_1^m)^j = 1^j = 1.$$

Combining our results for $m \mid j$ and $m \nmid j$, we finally get

$$\sum_{k=1}^{m}(1+\zeta_k)^n = m\cdot\sum_{k=0}^{\infty}\binom{n}{km}.$$

Now we will use trigonometric identities and de Moivre's formula to get another expression for the left side. First we use the double angle identities for cosine and sine to compute

$$\begin{aligned}1+\zeta_k = 1+e^{\frac{2k\pi}{m}i} &= \left(1+\cos\frac{2k\pi}{m}\right)+i\cdot\sin\frac{2k\pi}{m}\\ &= 2\cos^2\frac{k\pi}{m}+i\cdot 2\sin\frac{k\pi}{m}\cos\frac{k\pi}{m}\\ &= 2\cos\frac{k\pi}{m}\left(\cos\frac{k\pi}{m}+i\cdot\sin\frac{k\pi}{m}\right).\end{aligned}$$

By de Moivre's formula,

$$\sum_{k=1}^{m}(1+\zeta_k)^n = \sum_{k=1}^{m}\left[2^n\cdot\left(\cos\frac{k\pi}{m}\right)^n\cdot\left(\cos\frac{nk\pi}{m}+i\cdot\sin\frac{nk\pi}{m}\right)\right].$$

By equating this with our first expression of the same sum, we get

$$\sum_{k=0}^{\infty}\binom{n}{km}=\frac{2^n}{m}\cdot\sum_{k=1}^{m}\left(\cos\frac{k\pi}{m}\right)^n\cdot\left(\cos\frac{nk\pi}{m}+i\cdot\sin\frac{nk\pi}{m}\right).$$

The result follows from taking the real part of both sides because the left side is real and the imaginary part of the right side disappears.
The reader might be interested in knowing that there exists an even more general version of this theorem that evaluates

$$\sum_{k=0}^{\infty}\binom{n}{\ell+km},$$

where ℓ is a fixed non-negative integer. □

Sometimes, we have to resort to algebraic manipulations or induction, possibly in conjunction with applying known identities. Let us study this direct computation approach now.

Problem 5.24. Suppose n and m are positive integers and $k_1, k_2, \ldots, k_m$ are positive integers such that $k_1+k_2+\cdots+k_m=n$. Write the sum

$$\binom{n-1}{k_1-1, k_2, \ldots, k_m}+\binom{n-1}{k_1, k_2-1, \ldots, k_m}+\cdots+\binom{n-1}{k_1, k_2, \ldots, k_m-1}$$

as one multinomial coefficient. Deduce Pascal's identity (Corollary 4.6) as a consequence.

Example 5.25. Prove that, if n and m are integers such that $n > m \geq 0$, then

$$\sum_{k=0}^{m}(-1)^k\binom{n}{k}=(-1)^m\binom{n-1}{m}.$$

Unfortunately, there is no known closed form if we remove the alternating signs.

Solution. We will fix a positive integer n and perform induction on $m=0,1,2,\ldots,n-1$. In the base case $m=0$, the identity is easily seen to hold. Suppose the identity holds for some $m \geq 0$ and suppose $m+1$ is also less than or equal to $n-1$ (without the second condition, there is nothing left to prove). By the inductive hypothesis,

$$\begin{aligned}\sum_{k=0}^{m+1}(-1)^k\binom{n}{k}&=\sum_{k=0}^{m}(-1)^k\binom{n}{k}+(-1)^{m+1}\binom{n}{m+1}\\&=(-1)^m\binom{n-1}{m}+(-1)^{m+1}\binom{n}{m+1}.\end{aligned}$$

By Pascal's identity, this is equal to

$$(-1)^m\left[\binom{n-1}{m}-\binom{n}{m+1}\right]=(-1)^{m+1}\binom{n-1}{m+1},$$

which completes the induction. □

Problem 5.26. Prove, for all non-negative integers n, that

$$\sum_{k=0}^{n}\binom{n+k}{k}\cdot\frac{1}{2^k}=2^n.$$

Problem 5.27. Prove the following identities.

1. If n is a non-negative integer, then
$$\sum_{k=0}^{n} k\binom{n}{k} = n \cdot 2^{n-1}.$$
Use this identity to determine the sum of cardinalities of all elements of the power set of $[n] = \{1, 2, \ldots, n\}$ for non-negative integers n.

2. If n is a positive integer, then
$$\sum_{k=0}^{n} k^2\binom{n}{k} = \binom{n+1}{2} \cdot 2^{n-1}.$$

3. If n is a non-negative integer, then
$$\sum_{k=0}^{n} \frac{1}{k+1}\binom{n}{k} = \frac{2^{n+1}-1}{n+1}.$$

Example 5.28. If n and m are integers such that $n > m \geq 0$, show that
$$\sum_{k=m}^{n} (-1)^k \binom{n}{k}\binom{k}{m} = 0.$$

Solution. Using the identity
$$\binom{n}{k}\binom{k}{m} = \binom{n}{m}\binom{n-m}{k-m},$$
(see Problem 4.14), the sum becomes
$$\begin{aligned}\sum_{k=0}^{n} (-1)^k \binom{n}{k}\binom{k}{m} &= \binom{n}{m} \cdot \sum_{k=m}^{n} (-1)^k \binom{n-m}{k-m} \\ &= \binom{n}{m} \cdot \sum_{k=0}^{n-m} (-1)^{k+m} \binom{n-m}{k} \\ &= (-1)^m \binom{n}{m} \cdot \sum_{k=0}^{n-m} (-1)^k \binom{n-m}{k},\end{aligned}$$
where the final sum is equal to 0 by the binomial expansion of $(1-1)^{n-m}$. □

Problem 5.29. Prove the following identities.

1. If n and m are integers such that $n \geq m \geq 1$, then
$$\sum_{k=1}^{m} \binom{m}{k}\binom{n-1}{k-1} = \binom{m+n-1}{n}.$$

2. If n is a positive integer, then
$$\sum_{k=0}^{n} k\binom{n}{k}^2 = n \cdot \binom{2n-1}{n}.$$

3. For non-negative integers n, the nth Catalan number is $C_n = \frac{1}{n+1}\binom{2n}{n}$. Catalan numbers have hundreds of known combinatorial interpretations, some of which we will see in Section 10.3. For integers n and k such that $n \geq k \geq 1$, the Narayana number $N(n,k)$ is $\frac{1}{n}\binom{n}{k}\binom{n}{k-1}$. For all fixed positive integers n, show that
$$\sum_{k=1}^{n} N(n,k) = C_n.$$

Chapter 6

Inclusion-Exclusion

"I began at once somewhat more steady work on the subjects and books which I should have to lecture on. I now first hit upon the diagrammatical device of representing propositions by inclusive and exclusive circles. Of course the device was not new then, but it was so obviously representative of the way in which any one, who approached the subject from the mathematical side, would attempt to visualise propositions, that it was forced upon me almost at once."

– *John Venn, Symbolic Logic*

The addition principle (Theorem 1.31) tells us how to find the the cardinality of the union of finitely many disjoint finite sets in terms of the cardinalities of the constituent sets. The great restricting factor in this formula is the presumption of disjointedness. The principle of inclusion-exclusion tells us how to find the cardinality of such a union, even if the component sets are not disjoint. By alternating between adding (this is the inclusion) and subtracting (this is the exclusion) cardinalities of intersections of increasing depth among the component sets, we end up with a remarkable formula.

6.1 General Formulas

According to Theorem 1.37 and Theorem 1.40, if A, B, C are finite sets, then

$$|A \cup B| = |A| + |B| - |A \cap B|,$$
$$|A \cup B \cup C| = |A| + |B| + |C| - |A \cap B| - |B \cap C| - |C \cap A| + |A \cap B \cap C|.$$

The principle of inclusion-exclusion, PIE for short, generalizes these formulas to the union of n finite sets for every positive integer n. For the remainder of the section, let n be a positive integer and $A_1, A_2, \ldots, A_n$ be finite sets. As a reminder, we know from Corollary 1.38 that the union $A = \bigcup_{k=1}^{n} A_k$ is finite, and for every subset $J \subseteq [n] = \{1, 2, \ldots, n\}$, the intersection $\bigcap_{j \in J} A_j$ is finite. For each integer k such that $1 \leq k \leq n$, let

$$S_k = \sum_{\substack{J \subseteq [n] \\ |J| = k}} \left| \bigcap_{j \in J} A_j \right|,$$

where the sum is taken over all $\binom{n}{k}$ k-subsets of $[n]$. This allows us to state PIE and several variants, as follows.

Theorem 6.1 (Principle of inclusion-exclusion). The number of elements that lie in at least one of $A_1, A_2, \ldots, A_n$ is

$$\left| \bigcup_{k=1}^{n} A_k \right| = \sum_{k=1}^{n} (-1)^{k+1} S_k.$$

If the A_k are disjoint, then all of the terms in all of the S_k for $k > 1$ disappear, leaving us with the addition principle.

Proof. The idea is to write the cardinality of each set S as

$$|S| = \sum_{s \in S} 1.$$

Then it suffices to show that each element $a \in A$ contributes a total of 1 to each side of the stated equation. This is clearly the case on the left side. For the right side, we will do casework on the number of sets A_k in which a lies. For each integer m such that $1 \leq m \leq n$, let T_m be the sets of elements $a \in A$ such that a lies in exactly m of the A_k. For each $a \in A$, let $c(a)$ be the sum of the 1's and (-1)'s contributed by a on the right side. Then the right side is equal to

$$\sum_{k=1}^{n} (-1)^{k+1} S_k = \sum_{a \in A} c(a) = \sum_{m=1}^{n} \sum_{a \in T_m} c(a).$$

If $a \in T_m$, then a cannot lie in the intersection of more then m of the A_k; so a contributes nothing to S_k for $k > m$. For $k \leq m$, there are $\binom{m}{k}$ k-subsets of $\{A_1, A_2, \ldots, A_n\}$ in whose intersection a lies, by choosing k of the m A_k in which a lies. Using the binomial substitution method from Theorem 5.20, we find that

$$c(a) = \sum_{k=1}^{m} (-1)^{k+1} \binom{m}{k} = 1 - \sum_{k=0}^{m} (-1)^k \binom{m}{k} = 1 - (-1+1)^m = 1.$$

Therefore, the right side is $\sum_{m=1}^{n} \sum_{a \in T_m} 1$, which agrees with the left side $\sum_{a \in A} 1$. □

The formula for PIE may seem daunting due to the sheer number of terms and the alternating signs, but repeated usage will make it seem natural.

Corollary 6.2 (Symmetric PIE). Suppose it is true that, for each $k \in [n]$, there exists an integer α_k such that, for all k-subsets $J \subseteq [n]$,

$$\left| \bigcap_{j \in J} A_j \right| = \alpha_k.$$

Then PIE reduces to

$$\left| \bigcup_{k=1}^{n} A_k \right| - \sum_{k=1}^{n} (-1)^{k+1} \binom{n}{k} \alpha_k.$$

Proof. By the principle of inclusion-exclusion,

$$\begin{aligned} \left| \bigcup_{k=1}^{n} A_k \right| &= \sum_{k=1}^{n} (-1)^{k+1} \sum_{\substack{J \subseteq [n] \\ |J|=k}} \left| \bigcap_{j \in J} A_j \right| \\ &= \sum_{k=1}^{n} (-1)^{k+1} \sum_{\substack{J \subseteq [n] \\ |J|=k}} \alpha_k \\ &= \sum_{k=1}^{n} (-1)^{k+1} \binom{n}{k} \alpha_k, \end{aligned}$$

due to the fact that there are $\binom{n}{k}$ k-subsets $J \subseteq [n]$ for each $k = 1, 2, \ldots, n$. □

When proving the principle of inclusion-exclusion, we did casework on the exact number of A_k in which each element $a \in A$ lies. The next two results answer the finer questions of determining the number of $a \in A$ that lie in *exactly* or *at least* m of the A_k. Interestingly, it will turn out that only $S_m, S_{m+1}, \ldots, S_n$ are needed and $S_1, S_2, \ldots, S_{m-1}$ do not need to be a part of the equations. As in the proof of PIE, for the rest of the section, let T_m denote the set of elements $a \in A$ that inhabit exactly m of the A_k, for each integer m such that $1 \leq m \leq n$.

Theorem 6.3. For each integer m such that $1 \le m \le n$, the number of elements of A that lie in exactly m of the A_k is

$$|T_m| = \sum_{k=m}^{n} (-1)^{k+m} \binom{k}{m} S_k.$$

Proof. Suppose $a \in A$. As in the proof of PIE, let $c(a)$ denote the sum of the 1's and (-1)'s contributed by a to the right side. For each $a \in A$, there exists a unique integer t such that $1 \le t \le n$ and $a \in T_t$. Then the right side is equal to

$$\sum_{a \in A} c(a) = \sum_{t=1}^{n} \sum_{a \in T_t} c(a).$$

Fixing $a \in A$, if $t < m$, then $c(a) = 0$ because a cannot lie in the intersection of m or more of the A_k. If $t = m$, then a does not lie in the intersection of more than m of the A_k. So, a contributes to only the first term

$$(-1)^{m+m} \binom{m}{m} S_m = S_m$$

on the right side. Even for $t = m$, a lies in the intersection of exactly the $t = m$ of the A_k in which it lies, so $c(a) = 1$. Finally, if $t > m$, then a contributes

$$\sum_{k=m}^{t} (-1)^{k+m} \binom{k}{m} \binom{t}{k}$$

to the right side, where the upper bound on the index k gets cut off after t because a does not lie in the intersection of more than t of the A_k. By using Example 5.28, this sum evaluates to

$$(-1)^m \sum_{k=m}^{t} (-1)^k \binom{t}{k} \binom{k}{m} = 0.$$

Therefore, the right side is equal to

$$\sum_{t=1}^{n} \sum_{a \in T_t} c(a) = \sum_{a \in T_m} 1 = |T_m|.$$

As we can see, this method of counting $c(a)$ is applicable in the proof of more than one theorem. The caveat is that we must discover the correct formula at the beginning, usually through experimentation on small cases. □

Corollary 6.4. For each integer ℓ such that $1 \le \ell \le n$, the number of elements of A that lie in at least ℓ of the A_k is

$$\sum_{m=\ell}^{n} |T_m| = \sum_{k=\ell}^{n} (-1)^{k+\ell} \binom{k-1}{\ell-1} S_k.$$

Note that PIE is a special case of this formula when $\ell = 1$.

Proof. Using Theorem 6.3, the number of elements of A that lie in at least ℓ of the A_k is

$$\sum_{m=\ell}^{n} |T_m| = \sum_{m=\ell}^{n} \sum_{k=m}^{n} (-1)^{k+m} \binom{k}{m} S_k.$$

By a variation of the discrete Fubini's principle, we can swap the order of the indices to get that this is equal to

$$\sum_{k=\ell}^{n} \sum_{m=\ell}^{k} (-1)^{k+m} \binom{k}{m} S_k = \sum_{k=\ell}^{n} (-1)^k S_k \sum_{m=\ell}^{k} (-1)^m \binom{k}{m}.$$

The inner sum is the alternating sum of a tail end of a row of Pascal's triangle, which we can evaluate using Example 5.25 as

$$\begin{aligned}\sum_{m=\ell}^{k}(-1)^m\binom{k}{m} &= \sum_{m=0}^{k}(-1)^m\binom{k}{m} - \sum_{m=0}^{\ell-1}(-1)^m\binom{k}{m}\\ &= (-1+1)^k - (-1)^{\ell-1}\binom{k-1}{\ell-1}\\ &= (-1)^{\ell}\binom{k-1}{\ell-1}.\end{aligned}$$

The final expression is

$$\sum_{k=\ell}^{n}(-1)^{k+\ell}\binom{k-1}{\ell-1}S_k.$$

As a side note, if $\ell \leq n-1$, then complementary counting allows us to subtract the number of elements of A in at least $\ell+1$ of the A_k from the PIE expression for A to find the number of elements of A that are in *at most* ℓ of the A_k. Unfortunately, we were unable to simplify that resulting formula in a pretty way. □

The expression for PIE has many terms and it is natural to wonder whether we can estimate $|A|$ by dropping higher order terms. Bonferroni's inequalities state that this results in alternating between overestimating and underestimating $|A|$. The reader is asked to prove this in the following problem.

Problem 6.5 (Bonferroni's inequalities). Let n_1 be an odd integer in $[n]$ and n_2 be an even integer in $[n]$. Then

$$\sum_{k=1}^{n_2}(-1)^{k+1}S_k \leq \left|\bigcup_{k=1}^{n} A_k\right| \leq \sum_{k=1}^{n_1}(-1)^{k+1}S_k.$$

As a hint, it will be useful to use the $c(a)$ technique from the proof of PIE and its variants. Of course, if n_1 or n_2 is equal to n then equality holds in the corresponding inequality due to PIE. Also, note that for $n_1 = 1$, Bonferroni states the union bound (Problem 1.39), which is the combinatorial analogue of Boole's inequality from probability (Theorem 9.11).

In fact, all of the results that we have presented here have analogues in finite probability spaces. For example, a probabilistic version of PIE states: If n is a positive integer and $E_1, E_2, \ldots, E_n$ are events in a finite probability space, then

$$\mathbb{P}\left(\bigcup_{k=1}^{n} E_k\right) = \sum_{k=1}^{n}(-1)^{k+1}\sum_{\substack{J\subseteq[n]\\|J|=k}}\mathbb{P}\left(\bigcap_{j\in J} E_j\right).$$

The Bonferroni inequalities are actually usually stated in the context of probability. The proofs are not too different from the ones that we have shown in the solution. Instead of counting the sum of the (± 1)'s contributed by each $a \in A$, we count the number of atomic probabilities $(\pm p(\omega))$'s contributed by each $\omega \in \Omega$, where Ω is the sample space. For the reader's interest, we mention that there exists a general form of PIE for additive functions (briefly, these are real-valued functions that obey a law akin to Kolmogorov's third axiom (see Theorem 9.4)) that unifies the combinatorial and probabilistic versions of PIE, but this is not the place to discuss it.

Problem 6.6 (Maximum-minimums identity). Let n be a positive integer and $(x_i)_{i=1}^n$ be an n-tuple of real numbers. Prove that

$$\max(x_i)_{i=1}^n = \sum_{k=1}^{n}(-1)^{k+1}\sum_{\substack{J\subseteq[n]\\|J|=k}}\min(x_j)_{j\in J},$$

$$\min(x_i)_{i=1}^n = \sum_{k=1}^{n}(-1)^{k+1}\sum_{\substack{J\subseteq[n]\\|J|=k}}\max(x_j)_{j\in J}.$$

The reader who is familiar with number theory can use these to prove that, for all positive integers a, b, c,

$$\begin{aligned} \gcd(a,b)\cdot \operatorname{lcm}(ab) &= ab,\\ \operatorname{lcm}(a,b,c)\cdot\gcd(a,b)\cdot\gcd(b,c)\cdot\gcd(c,a) &= abc\cdot\gcd(a,b,c),\\ \gcd(a,b,c)\cdot\operatorname{lcm}(a,b)\cdot\operatorname{lcm}(b,c)\cdot\operatorname{lcm}(c,a) &= abc\cdot\operatorname{lcm}(a,b,c). \end{aligned}$$

6.2 Applications of PIE

There are several well-known problems to which PIE can be applied. We will explore some of these standard problems now. As an introductory note, we remark that it is often the case (possibly more often than not) that PIE is used in conjunction with complementary counting, because we wish to exclude several overlapping properties. All of our classic examples will fall in this category. Also, in most of our applications, it will turn out that we are working with the symmetric case of PIE (Corollary 6.2).

Theorem 6.7. Let n, s, t be positive integers and let the prime factorization of n be $n = p_1^{e_1}p_2^{e_2}\cdots p_m^{e_m}$. Then the number of s-tuples of positive integers $(q_1, q_2, \ldots, q_s) \in [t]^s$ such that $\gcd(n, q_1, q_2, \ldots, q_s) = 1$ is given by the function

$$J_s^t(n) = \sum_{k=0}^{m}(-1)^k \sum_{\substack{J\subseteq[m]\\|J|=k}} \left\lfloor \frac{t}{\prod_{j\in J} p_j}\right\rfloor^s.$$

Proof. We will use complementary counting. The total number of tuples in $[t]^s$ is t^s, so that is the number from which we will be subtracting. It is a fact from number theory that $\gcd(n, q_1, q_2, \ldots, q_s) = 1$ if and only if every prime that divides n does not divides some q_i. By contrapositive, $\gcd(n, q_1, q_2, \ldots, q_s) > 1$ if and only if there exists a prime factor of n that divides all of the q_i. For each $k \in [m]$, let T_k be the set of s-tuples

$$(q_1, q_2, \ldots, q_s) \in [t]^s$$

such that

$$p_k \mid q_1, p_k \mid q_2, \ldots, p_k \mid q_s.$$

Then we are seeking to compute

$$\begin{aligned} J_s^t(n) &= |\{q = (q_1, q_2, \ldots, q_s) \in [t]^s : \gcd(n, q_1, q_2, \ldots, q_s) = 1\}|\\ &= |\{q \in [n] : q \notin T_1 \wedge q \notin T_2 \wedge \cdots \wedge q \notin T_m\}|\\ &= t^s - |\{q \in [n] : q \in T_1 \vee q \in T_2 \vee \cdots \vee q \in T_m\}|\\ &= t^s - \left|\bigcup_{k=1}^{m} T_k\right| = t^s - \sum_{k=1}^{m}(-1)^{k+1}\sum_{\substack{J\subseteq[m]\\|J|=k}}\left|\bigcap_{j\in J} T_j\right|, \end{aligned}$$

where we used PIE to get to the last line. By a corollary of Bézout's lemma from Volume 3, if $\gcd(a, b) = 1$, then $a \mid c$ and $b \mid c$ if and only if $ab \mid c$. It is easy to generalize this by induction to any number of pairwise coprime divisors of c. Since distinct primes are pairwise coprime to each other, a ramification of this result is that each set $\bigcap_{j\in J} T_j$ consists of the s-tuples of multiples of $\prod_{j\in J} p_j$ in $[t]^s$. By a result in Volume 3, the number of multiples of a positive integer x in $[y]$, for a positive integer y, is $\left\lfloor \frac{y}{x}\right\rfloor$. So the number of s-tuples of the aforementioned kind is given by

$$\left|\bigcap_{j\in J} T_j\right| = \left\lfloor \frac{t}{\prod_{j\in J} p_j}\right\rfloor^s.$$

Finally, we can absorb the external term t^s into the sum as a zeroth term, which yields the desired formula. ☐

Definition 6.8. Given positive integers n and s, the sth **Jordan totient function** $J_s(n)$ counts the number of s-tuples $(q_1, q_2, \ldots, q_s) \in [n]^s$ such that $\gcd(n, q_1, q_2, \ldots, q_s) = 1$. For all positive integers n, **Euler's totient function** $\varphi(n)$ counts the number of integers $k \in [n]$ such that $\gcd(n, k) = 1$. Thus, the first Jordan totient function J_1 is equal to φ.

Corollary 6.9 (Jordan totient formula). Let n and s be positive integers, and let the prime factorization of n be

$$n = p_1^{e_2} p_2^{e_2} \cdots p_m^{e_m}.$$

Then the sth Jordan totient function may be evaluated as

$$J_s(n) = n^s \cdot \prod_{i=1}^{m} \left(1 - \frac{1}{p_i^s}\right).$$

Proof. This is a matter of factoring the formula in Theorem 6.7 for $t = n$. Since the product of any subset of the p_i divides n, our formula becomes

$$n^s + \sum_{k=1}^{m} (-1)^k \sum_{\substack{J \subseteq [m] \\ |J|=k}} \left| \bigcap_{j \in J} T_j \right| = n^s + \sum_{k=1}^{m} (-1)^k \sum_{1 \le i_1 < i_2 < \cdots < i_k \le m} \left(\frac{n}{p_{i_1} p_{i_2} \cdots p_{i_k}} \right)^s$$

$$= n^s \left(1 + \sum_{k=1}^{m} (-1)^k \sum_{1 \le i_1 < i_2 < \cdots < i_k \le m} \frac{1}{p_{i_1}^s p_{i_2}^s \cdots p_{i_k}^s} \right).$$

We would like to show that the expression inside the parentheses factors as

$$\left(1 - \frac{1}{p_1^s}\right)\left(1 - \frac{1}{p_2^s}\right) \cdots \left(1 - \frac{1}{p_m^s}\right).$$

It can be proven by induction on m that the expansion of this expression consists of 2^m terms, where each term is the product of m numbers, with one number from each of the factors $\left(1 - \frac{1}{p_k^s}\right)$. This results in the expansion

$$\sum_{k=0}^{m} \sum_{\substack{J \subseteq [m] \\ |J|=k}} \left[1^{m-k} \cdot \prod_{j \in J} \left(-\frac{1}{p_j^s} \right) \right] = 1 + \sum_{k=1}^{m} (-1)^k \sum_{\substack{J \subseteq [m] \\ |J|=k}} \prod_{j \in J} \frac{1}{p_j^s},$$

which is another way of writing what we wanted to see.
This generalizes Euler's totient function $\varphi = J_1$, which is given by the formula

$$\varphi(n) = n \cdot \prod_{k=1}^{m} \left(1 - \frac{1}{p_k}\right).$$

Euler's totient function will be studied in detail in Volume 3. □

Problem 6.10. Let s be a positive integer. It can be observed from the formula for the sth Jordan totient function J_s that it is a multiplicative arithmetic function. After becoming familiar with arithmetic functions (this is a topic of Volume 3), then prove that the summation function of J_s equals $S_{J_s}(n) = n^s$ for every positive integer n.

The formula for φ should be memorized, and it is easy to remember due to the symmetry among the primes in the formula, and because it is time-consuming to derive in the way shown on the spot. However, there is no need to memorize the next two formulas as they are fairly straightforward applications of PIE.

Theorem 6.11 (Derangements formula). Suppose n is a positive integer. A **derangement** of $[n]$ is a bijection $d : [n] \to [n]$ such that none of the elements $i \in [n]$ are a fixed point of d. As a reminder, a **fixed**

point of a function is an element of the domain that gets mapped to itself. The number of derangements of $[n]$ is

$$n! \cdot \sum_{k=0}^{n} \frac{(-1)^k}{k!}.$$

As a consequence, the probability of a bijection $f : [n] \to [n]$ being a derangement of $[n]$ approaches $\frac{1}{e}$ as n goes to infinity, where e is Euler's constant.

Proof. For each $k \in [n]$, let D_k be the set of bijections $f : [n] \to [n]$ such that k is a fixed point of f. For any positive integer m, the number of bijections $h : [m] \to [m]$ is $m!$ (Problem 3.14). By complementary counting and the symmetric variant of PIE, the number of derangements $d : [n] \to [n]$ is

$$\begin{aligned} n! - \left| \bigcup_{k=1}^{n} D_k \right| &= n! - \sum_{k=1}^{n} (-1)^{k+1} \sum_{\substack{J \subseteq [n] \\ |J|=k}} \left| \bigcap_{j \in J} D_j \right| \\ &= n! + \sum_{k=1}^{n} (-1)^k \sum_{\substack{J \subseteq [n] \\ |J|=k}} (n-k)! \\ &= n! + \sum_{k=1}^{n} (-1)^k \binom{n}{k} (n-k)! \\ &= n! + n! \cdot \sum_{k=1}^{n} \frac{(-1)^k}{k!} \\ &= n! \cdot \sum_{k=0}^{n} \frac{(-1)^k}{k!}, \end{aligned}$$

where we were able to absorb the $n!$ term into the sum as the 0th term. As a consequence, the probability of a bijection $f : [n] \to [n]$ being a derangement is

$$\frac{1}{n!} \cdot n! \cdot \sum_{k=0}^{n} \frac{(-1)^k}{k!} = \sum_{k=0}^{n} \frac{(-1)^k}{k!}.$$

Some knowledge of calculus tells us that the exponential function with base e can be expressed as $e^x = \sum_{k=0}^{\infty} \frac{x^k}{k!}$ for all real x. By substituting in $x = -1$, we get that $e^{-1} = \sum_{k=0}^{\infty} \frac{(-1)^k}{k!}$, which is exactly what the probability approaches as $n \to \infty$. □

Theorem 6.12. Suppose k, n, m are positive integers such that $n \geq k$. Then the number of functions $g : [m] \to [n]$, such that $[k]$ is a subset of the range of g, is

$$\sum_{i=0}^{k} (-1)^i \binom{k}{i} (n-i)^m.$$

Proof. For any positive integers r and s, we know that the total number of functions $f : [r] \to [s]$ is s^r. For each $i \in [n]$, let F_i be the number of functions $f : [m] \to [n]$ such that for all $\ell \in [m]$, $f(\ell) \neq i$; so f misses i. By complementary counting and the symmetric variant of PIE, the number of functions $g : [m] \to [n]$

such that $[k] \subseteq \operatorname{Rng}(g)$ is

$$\begin{aligned}
n^m - \left|\bigcup_{i=1}^{k} F_i\right| &= n^m - \sum_{i=1}^{k}(-1)^{i+1} \sum_{\substack{J\subseteq[k]\\|J|=i}} \left|\bigcap_{j\in J} F_j\right| \\
&= n^m + \sum_{i=1}^{k}(-1)^{i} \sum_{\substack{J\subseteq[k]\\|J|=i}} (n-i)^m \\
&= n^m + \sum_{i=1}^{k}(-1)^{i}\binom{k}{i}(n-i)^m \\
&= \sum_{i=0}^{k}(-1)^{i}\binom{k}{i}(n-i)^m,
\end{aligned}$$

where we could conveniently absorb n^m into the sum as the 0th term. □

Corollary 6.13 (Surjections formula)**.** Suppose m and n are positive integers. Then the number of surjections $g : [m] \to [n]$ is

$$\sum_{i=0}^{n}(-1)^{i}\binom{n}{i}(n-i)^m.$$

Proof. The formula is simply the instance of Theorem 6.12 when $k = n$, because then we are finding the the number of functions $g : [m] \to [n]$ such that g hits all of the elements of $[n]$ in its range. This result will appear in our study of distributions in Chapter 7 (specifically, Theorem 7.6), as a part of a larger network of problems. □

Corollary 6.14. Suppose k, n, m are positive integers such that $n \geq k$. If $m < k$, then

$$\sum_{i=0}^{k}(-1)^{i}\binom{k}{i}(n-i)^m = 0.$$

Proof. By Theorem 6.12,

$$\sum_{i=0}^{k}(-1)^{i}\binom{k}{i}(n-i)^m$$

counts the number of functions $g : [m] \to [n]$, such that $[k]$ is a subset of the range of g. However, since $m < k$, the reverse pigeonhole principle states that there are no such functions (Theorem 2.6). Thus, the expression is equal to 0. This identity will appear in our study of recurrence relations in Chapter 11 (specifically, Theorem 11.13) □

The following problem allows the reader to get some practice with using PIE, while also allowing us to foreshadow the introduction of graph theory in Chapter 8.

Problem 6.15. Informally, a graph (as in graph theory, and not the graph of a function) consists of dots called *vertices* on the plane and line segments called *edges* between pairs of vertices. The rules are that there are finitely many vertices, the two vertices of the endpoints of an edge are unordered (so the edge has no "direction"), every unordered pair of two distinct vertices has either 0 or 1 edges between them (but no more than 1), and no edge can have both of the vertices at its endpoints be the same vertex (so no "loops"). A vertex is said to be "isolated" if there are no edges emanating from it. For each positive integer n, determine the number of graphs on n distinguishable vertices such that none of the vertices are isolated.

Another general application of PIE will appear in the study of the κ function in Theorem 7.14.

Chapter 7

Distributions

> "There are actually formulas in the literature ("nameless here for evermore") for certain counting functions $f(n)$ whose evaluation requires listing all (or almost all) of the $f(n)$ objects being counted! Such a "formula" is completely worthless."
>
> – *Richard Stanley, Enumerative Combinatorics I*

We will now explore the problem of determining the number of ways of distributing n balls among k boxes. All of this material could be presented in a more formal way by formulating the statements in terms of special functions between finite sets, along with different notions of when two such functions are equivalent. We have avoided this extreme formalism in order to retain the intuition behind the ideas, but interested readers can research the "Twelvefold Way" in combinatorics [9].

7.1 General Problem and Easy Cases

Definition 7.1. Two objects or scenarios are said to be **distinguishable** if we can tell them apart, and are otherwise called **indistinguishable**. Another word for distinguishable is **distinct**.

Example. Two identical red balls are indistinguishable, but they become distinguishable if we paint the number 1 on one of them and the number 2 on the other.

Unless the problem is precisely stated, there is potential for ambiguity. In all cases of distributing balls among boxes, a specific ball can go into only one box. Several other factors that have to be clarified in a problem are:

1. Whether the balls are distinguishable or indistinguishable

2. Whether the boxes are distinguishable or indistinguishable

3. Whether there is a lower bound on the number of balls per box, such as there being at least one ball per box; unless otherwise specified, there is no lower bound, though we will usually be clear about whether empty boxes are allowed or not allowed

4. Whether there is an upper bound on the number of balls per box, such as there being at most one ball per box; unless otherwise specified, there is no upper bound

5. Whether there is a finite or infinite supply available of each distinguishable type of ball; unless otherwise specified, there will be a stated finite supply

6. Whether all of the balls are used up; unless otherwise specified, a distribution of a finite number of balls implies that each ball is placed in some box

7. Whether the order of the balls within each box matters; unless otherwise specified, the order of the balls within each box does not matter

The major results that will be derived in the ensuing sections are summarized below for the reader's convenience.

	n indistinguishable balls	n distinguishable balls
k distinguishable boxes, empty boxes allowed	$\binom{n+k-1}{k-1}$	k^n
k distinguishable boxes, no empty boxes	$\binom{n-1}{k-1}$	$\sum_{m=0}^{k}(-1)^m\binom{k}{m}(k-m)^n = k!\left\{ {n \atop k} \right\}$
k indistinguishable boxes, empty boxes allowed	$\sum_{i=1}^{\min(n,k)} \left\lvert {n \atop i} \right\rvert = \sum_{i=1}^{k} \left\lvert {n \atop i} \right\rvert$	$\sum_{i=1}^{\min(n,k)} \left\{ {n \atop i} \right\} = \sum_{i=1}^{k} \left\{ {n \atop i} \right\}$
k indistinguishable boxes, no empty boxes	$\left\lvert {n \atop k} \right\rvert$	$\left\{ {n \atop k} \right\}$

We will cover the most common general scenarios, including all of those in the table above, but we stress that it would be much more effective for the reader to understand the techniques than to memorize the formulas, as there are innumerable variations on the conditions that can be imposed in a problem on distributions. Any new notation will be defined at the necessary junctures. We will usually be working with a positive number of balls and a positive number of boxes, though results will be stated for a non-negative number of balls in some of the cases where empty boxes are allowed.

Problem 7.2. For positive integers $k \geq n$, find the number of ways of putting n distinguishable balls in k distinguishable boxes such that no box has more than one ball.

Problem 7.3. For positive integers $k \geq n$, find the number of ways of putting n indistinguishable balls in k distinguishable boxes such that no box has more than one ball.

The most straightforward scenarios are those of distributing n distinguishable balls to k distinguishable boxes. If empty boxes are allowed, then the problem is very easy. If empty boxes are not allowed, then the situation is a bit more complicated, but we can still derive a formula that allows for computation.

Theorem 7.4. For a non-negative integer n and a positive integer k, the number of ways of distributing n distinguishable balls to k distinguishable boxes, if empty boxes are allowed, is k^n.

Proof. For each of the n distinguishable balls, there are k choices of distinguishable boxes in which it can be placed. By the multiplication principle, the answer is $\underbrace{k \cdot k \cdots k}_{n \text{ copies of } k} = k^n$. □

Problem 7.5. Let $n \geq k$ be positive integers. There are n tubs of k balls each, such that balls within the same tub are indistinguishable but balls from different tubs are distinguishable. There are k distinguishable boxes. In how many ways can exactly one ball be placed in each box?

Theorem 7.6. Let n and k be positive integers such that $n \geq k$. Then the number of ways of distributing n distinguishable balls to k distinguishable boxes, where empty boxes are not allowed, is

$$\sum_{m=0}^{k}(-1)^m\binom{k}{m}(k-m)^n.$$

Proof. The number of such distributions is the number of surjections $g : [n] \to [k]$. We found a formula for counting surjections in Corollary 6.13. □

7.2 Compositions

Now we will consider the scenario of distributing n indistinguishable balls to k distinguishable boxes, along with related problems. This case is the one that is most easily computed as it boils down to binomial coefficients. The argument is not easy to independently develop though, as we will require a clever bijection.

Definition 7.7. For non-negative integers n and k, then we say that a **weak k-composition** of n is a list $(a_1, a_2, \ldots, a_k)$ of k non-negative integers such that

$$a_1 + a_2 + \cdots + a_k = n.$$

If the a_i are restricted to being positive, then we call it a **k-composition** of n. The a_i are called the **components** of a (weak) composition. (Lists were defined in Definition 1.17.)

Theorem 7.8 (Weak compositions). For a non-negative integer n and a positive integer k, the following problems all lead to the same answer, which is $\binom{n+k-1}{k-1}$.

1. Find the number of ways in which n indistinguishable balls can be distributed to k distinguishable boxes, such that empty boxes are allowed.
2. Find the number of weak k-compositions of n.
3. Find the number of lists $(a_1, a_2, \ldots, a_{k-1})$ of $k-1$ non-negative integers such that

   $$a_1 + a_2 + \cdots + a_{k-1} \leq n.$$

4. Find the number of ways of constructing an n-multinomial set, given a fixed support that is a set of k elements. Note that, since the k elements form a set, they are distinct. Multinomial sets were defined in Definition 3.36, where it was said that the multiplicity of each of the k elements is non-negative.

Proof. For $n = 0$, all the problems logically lead to a result of $\binom{n+k-1}{k-1} = 1$, so we can assume that n is positive, which will make the arguments easier to conceptualize. We will use a method that has three names: balls and urns, stars and bars, and sticks and stones, the last being the most relevant to how we will phrase our solution.

1. Place n balls (i.e. the stones) in a row. Then add $k-1$ indistinguishable dividers (i.e. the sticks) at the right end of the row. Each permutation of these $k-1$ sticks and n stones will produce k sections, as created by consecutive sticks. That is, there is a section to the left of the leftmost stick, a section in between the leftmost and second-leftmost stick, and so on until we reach the section to the right of the rightmost stick. To be clear, in some permutations, there will be sections with no stones. Here is a visualization for one way of distributing 5 ball across 4 boxes (so there are 3 sticks):

   $$\circ \mid \circ \circ \circ \mid \circ \mid$$

   If we label the sections as $S_1, S_2, \ldots, S_k$ then the number of stones in S_i corresponds exactly with the number of balls in box i. Each such permutation leads to a unique way of distributing n balls to k boxes and each way of distributing n balls to k boxes leads back to a unique such permutation. Thus, we have a bijection, and the answer is the number of such permutations, which is

   $$\frac{(n+k-1)!}{n!(k-1)!} = \binom{n+k-1}{k-1}.$$

2. The weak k-compositions of n are in bijection with the ways of distributing n indistinguishable balls to k distinguishable boxes. This is because a k-composition

   $$(a_1, a_2, \ldots, a_k)$$

   of n can be interpreted as the unique way of placing a_i balls into box i, for each i. The inverse map is also injective, so we have a bijection.

3. The clever idea here is to associate each list

$$(a_1, a_2, \ldots, a_{k-1})$$

such that

$$a_1 + a_2 + \cdots + a_{k-1} \leq n$$

with the list $(a_1, a_2, \ldots, a_k)$ where we define the extra component as

$$a_k = n - (a_1 + a_2 + \cdots + a_{k-1}).$$

Note that $a_k \geq 0$ and

$$a_1 + a_2 + \cdots + a_k = n.$$

This produces a bijection, so it is equivalent to count the number of weak k-compositions of n.

4. Let the set of k elements be $\{b_1, b_2, \ldots, b_k\}$. In a multinomial set with n elements chosen from among the b_i, let a_i be the number of times that b_i appears for each i. Then

$$a_1 + a_2 + \cdots + a_k = n.$$

This is a weak k-composition of n and each n-multinomial set leads to a unique one, giving an injection. In the other direction, any weak k-composition of n can be used to reconstruct a unique n-multinomial set, meaning we also have an injection in the other direction. So, it is equivalent to find the number of weak k-compositions of n.

Thus, we have shown that all of these problems are equivalent. □

Definition 7.9. For non-negative integers n and positive integers k, the numbers of the form $\binom{n+k-1}{k-1}$ are called **multiset coefficients** and are denoted by $\left(\!\binom{k}{n}\!\right)$. This is a touch confusing, as our terminology and Theorem 7.8 would imply that we should call these "multinomial set coefficients" instead. In any case, we will not see this notation much because binomial coefficients suffice.

Example 7.10. For a non-negative integer n and a positive integer k, it holds that

$$\left(\!\binom{k}{n}\!\right) = \left(\!\binom{n+1}{k-1}\!\right).$$

Solution. The symmetric property $\binom{n}{k} = \binom{n}{n-k}$ of binomial coefficients gives

$$\left(\!\binom{k}{n}\!\right) = \binom{n+k-1}{k-1} = \binom{n+k-1}{n} = \binom{(k-1)+(n+1)-1}{(n+1)-1} = \left(\!\binom{n+1}{k-1}\!\right).$$

Strangely, this means that the number of ways of distributing n indistinguishable balls across k distinguishable boxes is equal to the number of ways of distributing $k-1$ indistinguishable balls across across $n+1$ distinguishable boxes. A combinatorial interpretation that allows us to understand this phenomenon is that we can interpret sticks as stones and stones as sticks in the bijection from Theorem 7.8. □

Problem 7.11. Let m and n be positive integers such that $m \geq n$. Determine the number of strictly increasing functions $f : [n] \to [m]$, and, separately, the number of non-decreasing functions $g : [n] \to [m]$.

Problem 7.12. Recall that the multinomial theorem (Theorem 5.6) says that, for any positive integer m and any non-negative integer n,

$$(x_1 + x_2 + \cdots + x_m)^n = \sum_{\substack{k_1+k_2+\cdots+k_m=n \\ k_i \text{ non-negative}}} \binom{n}{k_1, k_2, \ldots, k_m} x_1^{k_1} x_2^{k_2} \cdots x_m^{k_m}.$$

Determine the number of terms in this sum.

Problem 7.13. Let n be a positive integer with prime factorization $n = p_1^{e_1} p_2^{e_2} \cdots p_m^{e_m}$ where the p_i are distinct primes. Produce a formula that counts the number of k-tuples of positive integers $(a_1, a_2, \ldots, a_k)$ such that

$$a_1 a_2 \cdots a_k = n.$$

What if the a_j are allowed to be negative? As a side note, this is an interesting problem because it determines the number of cases that need to be addressed when solving a Diophantine equation by "fudging and factoring," which is a technique that we will study in Volume 3.

Theorem 7.14. Let n be a non-negative integer, k be a positive integer, and m be a positive integer. The number of weak k-compositions of n such that no component exceeds $m-1$ is denoted by and computed as

$$\kappa(n,k,m) = \sum_{i=0}^{k} (-1)^i \binom{k}{i} \binom{n - mi + k - 1}{k-1}.$$

We use the convention that $\binom{a}{b} = 0$ if $a < b$. As a throwback to generating functions, this is the coefficient of x^n in the expansion of

$$(1 + x + x^2 + \cdots + x^{m-1})^k.$$

Proof. For each $i \in [k]$, we use A_i to denote the set of weak k-compositions of n,

$$a_1 + a_2 + \cdots + a_k = n,$$

such that $a_i \geq m$ (and so $a_i > m-1$). By complementary counting and PIE, along with sticks and stones, we are seeking a formula for

$$\binom{n+k-1}{k-1} - \left|\bigcup_{i=1}^{k} A_i\right| = \binom{n+k-1}{k-1} - \sum_{i=1}^{k} (-1)^{i+1} \sum_{\substack{J \subseteq [k] \\ |J| = i}} \left|\bigcap_{j \in J} A_j\right|.$$

We will now compute the inner summands. Fix any $i \in [k]$ and then $J \subseteq [k]$ such that $|J| = i$. Let $a_1 + a_2 + \cdots + a_k = n$ be a weak k-composition of n such that if $j \in J$ then a_j exceeds m. We can subtract mi from both sides of the equation to get

$$\sum_{\ell \notin J} a_\ell + \sum_{j \in J} (a_j - m) = n - mi,$$

which is a weak k-composition of $n - mi$. This map is easily verified to be a bijection. So, by sticks and stones,

$$\sum_{\substack{J \subseteq [k] \\ |J| = i}} \left|\bigcap_{j \in J} A_j\right| = \binom{k}{i} \binom{n - mi + k - 1}{k-1},$$

where the first factor follows from the fact that there are $\binom{k}{i}$ ways to build J. This combinatorial argument makes sense if and only if i satisfies $n - mi \geq 0$, but the formula still works if $n < mi$ because those terms are 0, as they combinatorially should be. Thus, the final formula is

$$\binom{n+k-1}{k-1} - \sum_{i=1}^{k} (-1)^{i+1} \binom{k}{i} \binom{n - mi + k - 1}{k-1},$$

which is equal to the sum in the statement of the theorem because the term external to the sum can be absorbed into the sum as the 0th term. We leave it to the reader to become convinced of the note on the generating function. □

Problem 7.15. In the notation of Theorem 7.14, prove that

$$\sum_{n=0}^{(m-1)k} \kappa(n,k,m) = m^k.$$

As a hint, this can be done via double counting and without any algebra.

Now we will see an analogue of sticks and stones, where there is an "at least" restriction in each of the problems.

Theorem 7.16 (Compositions)**.** For positive integers n and k such that $n \geq k$, the following problems all lead to the same answer, which is the binomial coefficient $\binom{n-1}{k-1}$.

1. Find the number of ways in which n indistinguishable balls can be distributed to k distinguishable boxes, such that empty boxes are not allowed, meaning each box receives at least one ball.
2. Find the number of k-compositions of n.
3. Find the number of lists $(a_1, a_2, \ldots, a_{k-1})$ of $k-1$ positive integers such that
$$a_1 + a_2 + \cdots + a_{k-1} \leq n-1.$$
4. Find the number of ways of constructing an n-multiset, given a fixed support that is a set of k elements. Multisets were defined in Definition 1.21, where it was stated that the multiplicity of each of the k elements must be positive.

Proof. For $n = k$, all the problems logically lead to a result of $\binom{n-1}{k-1} = 1$, so we can assume that $n > k$, which will make the arguments easier to conceptualize. We will show that the solution to each problem in this list follows from the solution to the analogous problem in Theorem 7.8, except here we have an "at least" condition.

1. Suppose n indistinguishable balls are distributed to k distinguishable boxes in a way that each box has at least one ball. If we remove one ball from each box in such a distribution, then we get a unique way of putting $n-k$ of the balls into k boxes with possibly empty boxes. In the other direction, for each way of putting $n-k$ of the balls into k boxes with possibly empty boxes, adding a ball to each box gives a unique distribution of n balls in k boxes with at least one ball in each box. So we have a bijection, and the answer, using the analogous problem where empty boxes are allowed, is
$$\binom{(n-k)+k-1}{k-1} = \binom{n-1}{k-1}.$$
2. The k-compositions of n are in bijection with the weak k-compositions of $n-k$. This is because we can take a k-composition $(a_1, a_2, \ldots, a_k)$ of n and subtract 1 from each a_i to get the equation
$$(a_1 - 1) + (a_2 - 1) + \cdots + (a_k - 1) = n - k,$$
and so we have produced a unique weak k-composition $(a_1 - 1, a_2 - 1, \ldots, a_k - 1)$ of $n-k$. The reverse operation works as well, and we get a bijection. So, it is equivalent to count the number of weak k-compositions of $n-k$, which produces the desired answer.
3. The same technique as in the previous part works. We simply transition between
$$a_1 + a_2 + \cdots + a_{k-1} \leq n-1$$
and
$$(a_1 - 1) + (a_2 - 1) + \cdots + (a_{k-1} - 1) \leq (n-1) - (k-1) = n-k.$$
Due to this bijection, it is equivalent to count the number of lists $(c_1, c_2, \ldots, c_{k-1})$ of $k-1$ non-negative integers such that
$$c_1 + c_2 + \cdots + c_{k-1} \leq n-k.$$
Using the formula from Theorem 7.8, we get the answer.

4. If we have an n-multiset with a support of k elements $\{b_1, b_2, \ldots, b_k\}$, where each b_i appears at least once in the n-multiset, then we can remove one copy of each b_i from the n-multiset to produce a unique $(n-k)$-multinomial set with a non-negative number of copies of each b_i. The reverse operation works as well, so this is bijection. Thus, it is equivalent to count the number of $(n-k)$-multinomial sets with a non-negative number of copies of each b_i, which produces the expected answer.

Thus, we have shown that all of these problems are equivalent. □

Problem 7.17. Prove that

$$\left(\!\!\binom{n}{k-1}\!\!\right) + \left(\!\!\binom{n-1}{k}\!\!\right) = \left(\!\!\binom{n}{k}\!\!\right).$$

Example 7.18. For positive integers n and k such that $n \geq k$, find the number of ways of distributing n distinguishable balls to k indistinguishable boxes such that empty boxes are not allowed and the order of the balls within each box matters.

Solution. A common problem-solving technique is to temporarily change an undesirable condition to a desirable one. Let us start with instead finding the number of ways of distributing n distinguishable balls to k *distinguishable* boxes such that each box gets at least one ball and the order of the balls within each box matters; call the set of such distributions S. We assert that the number of such distributions is in bijection with the Cartesian product $P \times D$ where P is the collection of permutations of n distinguishable balls, and D is the collection of ways of distributing n indistinguishable balls to k distinguishable boxes such that each box gets at least one ball. Here is why:
Say we have an element of S. After distributing the balls, if we number the boxes and place them in a row, this produces a permutation of the balls since the balls in each box are also in a row; and if we interpret the balls as being indistinguishable, this also induces an element of D. This map from S to $P \times D$ is injective. Conversely, if we have an element of P and an element of D, we can injectively construct an element of S by putting the boxes in a row and replacing their indistinguishable balls one by one with the distinguishable balls in the order of the permutation. So a bijection exists and

$$|S| = |P \times D| = |P| \cdot |D| = n!\binom{n-1}{k-1}.$$

Finally, we switch from the distinguishable boxes of S to the indistinguishable boxes of the original problem by noting that each permutation of boxes appears $k!$ times. So only 1 out of every $k!$ permutations of the boxes is distinct, giving a final answer of

$$L(n,k) = \frac{n!}{k!}\binom{n-1}{k-1}.$$

These $L(n,k)$ are called **Lah numbers**. As a side note, suppose we had instead been asked for the number of ways of distributing n distinguishable balls to k indistinguishable boxes such that *empty boxes are allowed* and the order of the balls within each box matters. Then the answer would have been

$$\sum_{i=1}^{k} L(n,i) = L(n,1) + L(n,2) + \cdots + L(n,k),$$

because each summand $L(n,i)$ stands for the case of exactly $k-i$ empty boxes and exactly i boxes with at least one ball. We do not believe that a closed form exists for this sum. In particular, the same argument does not work because one empty box cannot be distinguished from another. □

Problem 7.19. For a non-negative integer n and positive integer k, find the number of ways of distributing n distinguishable balls to k distinguishable boxes, if empty boxes are allowed, and the order of the balls within each box matters.

Problem 7.20. For a positive integer k and integers $p \leq q$, find the number of lists $(c_1, c_2, \ldots, c_k)$ of k integers such that

$$p \leq c_1 \leq c_2 \leq \cdots \leq c_k \leq q.$$

Example 7.21. Let k be a fixed positive integer. Prove that the generating functions for weak k-compositions and for k-compositions converge on an interval around 0, and find closed forms for them. Here, it is the number of balls $n \geq 0$ that varies. So that this problem makes sense, we define that the k-composition $\binom{n-1}{k-1}$ equals 0 if $n < k$, including the case $n = 0$.

Solution. The generating function for weak k-compositions equals

$$\sum_{n=0}^{\infty} \binom{n+k-1}{k-1} x^n = (1 + x + x^2 + x^3 + \cdots)^k$$

because, by a one-to-one correspondence, the coefficient of each x^n in both expressions counts the number of weak k-compositions of n. As a function, the theorem on infinite geometric series tells us that the factors of the latter are geometric series that converge for $|x| < 1$. Thus,

$$\sum_{n=0}^{\infty} \binom{n+k-1}{k-1} x^n = \left(\sum_{i=0}^{\infty} x^i\right)^k = \frac{1}{(1-x)^k}.$$

Using a similar correspondence, the generating function for k-compositions is

$$\sum_{n=0}^{\infty} \binom{n-1}{k-1} x^n = (x + x^2 + x^3 + \cdots)^k = x^k \cdot \left(\sum_{i=0}^{\infty} x^i\right)^k,$$

which converges to

$$x^k \cdot \frac{1}{(1-x)^k} = \left(\frac{x}{1-x}\right)^k$$

for $|x| < 1$. So, the interval $(-1, 1)$ suffices in either case. □

7.3 Partitions

We have previously seen techniques for when there are distinguishable boxes. In this section, we will work with indistinguishable boxes, both for distinguishable balls and indistinguishable balls.

Definition 7.22. If S is a non-empty set, then a **set partition** of S is a set of non-empty disjoint subsets of S such that the union of these subsets is S; this was also stated as a part of Definition 1.30. For each positive integer n, the total number of partitions of a set of n elements is called the **Bell number** B_n.

Example. The following are the partitions of the set $\{a, b, c\}$, ordered from the least number of elements to the most:

$$\{\{a\}, \{b\}, \{c\}\}, \quad \{\{a\}, \{b, c\}\}, \{\{b\}, \{a, c\}\}, \{\{c\}, \{a, b\}\}, \quad \{\{a, b, c\}\}.$$

Thus, we have found the Bell number $B_3 = 5$.

Definition 7.23. If a set partition has k sets in it, then we call it a **k-partition**. For positive integers n and k such that $n \geq k$, the **Stirling partition number** $\left\{ {n \atop k} \right\}$ is the number of k-partitions of a set of n elements. If $n < k$, we define $\left\{ {n \atop k} \right\} = 0$ for ease of notation and to match the combinatorial meaning of this degenerate concept. Note that

$$B_n = \sum_{k=1}^{n} \left\{ {n \atop k} \right\}.$$

Problem 7.24. For positive integers n, find explicit formulas for

$$\left\{ {n \atop 1} \right\}, \left\{ {n \atop n} \right\}, \left\{ {n \atop 2} \right\}, \left\{ {n \atop n-1} \right\}.$$

Problem 7.25. Prove that the Stirling partition numbers (Definition 7.23) satisfy the following identity for positive integers n and k such that $n \geq k+1$:

$$\left\{ {n+1 \atop k+1} \right\} = \left\{ {n \atop k} \right\} + (k+1) \left\{ {n \atop k+1} \right\}.$$

Theorem 7.26. Let $n \geq k$ be positive integers. The Stirling partition number $\left\{ {n \atop k} \right\}$ is equal to the number of ways of distributing n distinguishable balls to k indistinguishable boxes, where no box is allowed to be empty. This number evaluates to

$$\frac{1}{k!} \sum_{i=0}^{k} (-1)^i \binom{k}{i} (k-i)^n.$$

Proof. The number of k-partitions of a set of n elements is easily seen to be in bijection with the set of ways of distributing n distinguishable balls to k indistinguishable boxes, where no box is allowed to be empty. So, our job is now to prove the formula. This is also easy because we know from Theorem 7.6 that the number of ways of distributing n distinguishable balls to k *distinguishable* boxes, where no box is allowed to be empty, is

$$\sum_{i=0}^{k} (-1)^i \binom{k}{i} (k-i)^n.$$

We wish to switch from distinguishable boxes to indistinguishable boxes. Then we simply have to note that distributions with distinguishable boxes split into disjoint classes of $k!$ distributions, with each class consisting of permutations of each other. The set of those classes are in bijection with distributions with indistinguishable boxes. Thus, we divide by $k!$ to get the answer. □

Corollary 7.27. Let $n \geq k$ be positive integers. Using the notation of Stirling partition numbers, the number of ways of distributing n distinguishable balls to k distinguishable boxes, where empty boxes are not allowed, is $k!\left\{ {n \atop k} \right\}$.

Corollary 7.28. Let n and k be positive integers. The number of ways of distributing n distinguishable balls to k indistinguishable boxes, where boxes are allowed to be empty, is

$$\sum_{i=1}^{\min(n,k)} \left\{ {n \atop i} \right\} = \sum_{i=1}^{k} \left\{ {n \atop i} \right\}.$$

Proof. This is a matter of casework and using the addition principle. We start by interpreting the left side. For each index i, the summand $\left\{ {n \atop i} \right\}$ stands for the case of exactly i boxes with at least one ball and exactly $k-i$ empty boxes. If $n \geq k$, there are anywhere from 1 to k, inclusive, boxes that can have at least one ball. If $n < k$, there are anywhere from 1 to n, inclusive, boxes that can have at least one ball. Thus, the upper bound of $\min(n,k)$ on the index is correct. On the right side, an upper bound of k on the index is also correct because we defined $\left\{ {n \atop i} \right\}$ to equal 0 for $i > n$. □

Now, we will explore the final and least computation-friendly case, that of distributing indistinguishable balls to indistinguishable boxes.

Definition 7.29. Recall from Definition 5.14 that a **partition of an integer** $n \geq 1$ is a multiset of positive integers whose sum is n. (As a reminder, multisets can have repeated elements, and order does not matter in multisets.) The total number of partitions of n is denoted by $p(n)$ and this is the called the **partition function**.

Example. The following are the partitions of 5, ordered from the least number of elements to the most:

$$\langle 5\rangle,\ \ \langle 4,1\rangle,\langle 3,2\rangle,\ \ \langle 3,1,1\rangle,\langle 2,2,1\rangle,\ \ \langle 2,1,1,1\rangle,\ \ \langle 1,1,1,1,1\rangle.$$

Thus, we have found that $p(5) = 7$.

Definition 7.30. If a partition of n has $k \geq 1$ elements, then we call it a k-**partition**. The number of k-partitions of n is denoted by $\left|\begin{matrix} n \\ k \end{matrix}\right|$ or $p_k(n)$. Partitions make sense for $n \geq k$, but if $n < k$, we define $\left|\begin{matrix} n \\ k \end{matrix}\right| = 0$ for ease of notation and to match the combinatorial meaning of this degenerate concept, similar to set partitions. Note that

$$p(n) = \sum_{k=1}^{n} \left|\begin{matrix} n \\ k \end{matrix}\right|.$$

No general closed formula is known for $p(n)$ or $\left|\begin{matrix} n \\ k \end{matrix}\right|$.

Problem 7.31. Compute $\left|\begin{matrix} n \\ 1 \end{matrix}\right|$, $\left|\begin{matrix} n \\ n \end{matrix}\right|$, and $\left|\begin{matrix} n \\ 2 \end{matrix}\right|$ for integers $n \geq 1$, and compute $\left|\begin{matrix} n \\ n-1 \end{matrix}\right|$ for integers $n \geq 2$.

Problem 7.32. Prove that partitions of integers satisfy the following identity for n and k such that each term is defined:

$$\left|\begin{matrix} n \\ k \end{matrix}\right| = \left|\begin{matrix} n-1 \\ k-1 \end{matrix}\right| + \left|\begin{matrix} n-k \\ k \end{matrix}\right|.$$

Theorem 7.33. Let $n \geq k$ be positive integers. The number of ways of distributing n indistinguishable balls to k indistinguishable boxes, where no box is allowed to be empty, is $\left|\begin{matrix} n \\ k \end{matrix}\right|$.

Proof. We will construct a bijection between the types of distributions described and the k-partitions of n. Given such a distribution, we can interpret the number of balls in each box as an element of a k-partition of n. Since there are n balls in total, this does produce a k-partition of n. The distribution can be recovered from the k-partition because we can simply interpret each element a in the k-partition as a box with a balls. So the map is injective. Moreover, this reconstruction of a distribution using a k-partition can be done with any k-partition, so the map is surjective. Thus, there is a bijection. □

Corollary 7.34. Let n and k be positive integers. The number of ways of distributing n indistinguishable balls to k indistinguishable boxes, where boxes are allowed to be empty, is

$$\sum_{i=1}^{\min(n,k)} \left|\begin{matrix} n \\ i \end{matrix}\right| = \sum_{i=1}^{k} \left|\begin{matrix} n \\ i \end{matrix}\right|.$$

Proof. As evident from the format of the formula, the derivation will be very similar to the case of putting n distinguishable balls into k indistinguishable boxes, where boxes are allowed to be empty (Corollary 7.28). We will apply the addition principle. We start by interpreting the left side. For each index i, the summand $\left|\begin{matrix} n \\ i \end{matrix}\right|$ stands for the case of exactly i boxes with at least one ball and exactly $k-i$ empty boxes. If $n \geq k$, there are anywhere from 1 to k, inclusive, boxes that can have at least one ball. If $n < k$, there are anywhere from 1 to n, inclusive, boxes that can have at least one ball. Thus, the upper bound of $\min(n, k)$ on the index is correct. On the right side, an upper bound of k on the index is also correct because we defined $\left|\begin{matrix} n \\ i \end{matrix}\right|$ to equal 0 for $i > n$. □

Chapter 8

Graph Theory

"When drawing a graph on a piece of paper, we naturally try to do this as transparently as possible. One obvious way to limit the mess created by all the lines is to avoid intersections."

– *Reinhard Diestel, Graph Theory*

"Suppose aliens invade the earth and threaten to obliterate it in a year's time unless human beings can find the Ramsey number for red five and blue five. We could marshal the world's best minds and fastest computers, and within a year we could probably calculate the value. If the aliens demanded the Ramsey number for red six and blue six, however, we would have no choice but to launch a preemptive attack."

– *Paul Erdős, Scientific American*

Graph theory is not about graphing functions. Visually, it is about dots, some pairs of which are connected by curves. This is a representation that turns out to be useful in many contexts, such as computer science. The subject matter is vast with far-reaching applications, so we will inevitably skip many interesting topics, such as Hamiltonian paths. After studying some basic ideas, including trees, we will follow up with observations about planar graphs, which have applications, for example, in geometry. Finally, we will look at an area called Ramsey theory, which begins by finding regularities in edge colourings of sufficiently large graphs.

8.1 Trees

Definition 8.1. A **graph** G consists of a finite set $\mathcal{V}$ of elements called **vertices**, and a set $\mathcal{E}$ of elements called **edges**, where each edge is a set of two vertices (an edge $\{v_1, v_2\}$ may be written as v_1v_2 for convenience of notation). As a reminder, sets cannot have repeated elements and order does not matter in sets. Visually, vertices can be thought to be distinct points in the plane and edges can be thought to be line segments or curves connecting pairs of distinct vertices. Unlike in polygons, we do not include the endpoints in an edge, in order for vertices to be disjoint from edges. We will assume that a graph has a non-empty set of vertices.

Definition 8.2. There are several standard ways in which aspects of the definition of a graph can be altered to allow for other structures:

- Whereas graphs have a finite number of vertices and subsequently a finite number of edges, an **infinite graph** has an infinite number of vertices and potentially an infinite number of edges. To emphasize that this is not allowed in a graph, we will call an ordinary graph **finite**.

- Whereas the collection of edges of a graph forms a set, the collection of edges of a **multigraph** forms a multiset. This means there can be multiple edges between two unordered vertices. To emphasize that this is not allowed in a graph, we will call an ordinary graph **simple**.

- Whereas each edge of a graph forms a set of two elements, the edges of a **directed graph** are lists of two distinct elements and so are ordered. Visually, we would draw an arrow from the first vertex to the second vertex to represent the order. Depending on the author, both (x, y) and (y, x) may or may not be allowed to be edges. To emphasize that directed edges are not allowed in a graph, we will call an ordinary graph **undirected**.

- Whereas each edge of a graph forms a set of two elements, the edges of a **graph with loops** are multisets of two elements and so allow for edges between the same one vertex. Visually, we would draw an edge or loop from one vertex to itself. To emphasize that this is not allowed in a graph, we will call an ordinary graph a **graph without loops**.

Combinations of these deviations form further structures. Unless otherwise specified, our graphs are finite, simple, undirected, and without loops.

Problem 8.3. A **complete graph** K_n on n vertices has one edge between every unordered pair of vertices. How many edges does K_n have?

Definition 8.4. The **degree of a vertex** v is the number of edges emanating from it, and it is denoted by $\deg v$. If $\deg v = 1$ for some vertex v, then we call v a **leaf**.

Theorem 8.5 (Handshaking lemma). The following two results are interchangeably called the handshaking lemma, but we will only refer to the first as such, in order to avoid ambiguity.

1. Let $\mathcal{V}$ be the set of vertices of a graph and $\mathcal{E}$ be the set of edges. Then

$$\sum_{v\in\mathcal{V}} \deg v = 2|\mathcal{E}|.$$

2. Every graph has an even number of vertices of odd degree.

Proof. The second part will follow from the first part:

1. We will use a double counting argument due to Euler. Let P be the set of pairs (v, e) such that $v \in \mathcal{V}$ and $e \in \mathcal{E}$ and v is a vertex at an endpoint of e. The two methods are casework on $v \in \mathcal{V}$ and casework on $e \in \mathcal{E}$. For each $v \in \mathcal{V}$, there are $\deg v$ possible $e \in \mathcal{E}$ with which v could be paired. On the other hand, for each $e \in \mathcal{E}$, there are 2 possible $v \in \mathcal{V}$ with which e could be paired. Therefore,

$$\sum_{v\in\mathcal{V}} \deg v = |P| = 2|\mathcal{E}|.$$

2. Let $\mathcal{V}_1$ be the set of vertices in $\mathcal{V}$ of odd degree, and let $\mathcal{V}_2$ be the set of vertices in $\mathcal{V}$ of even degree. By the handshaking lemma,

$$2|\mathcal{E}| = \sum_{v\in\mathcal{V}} \deg v = \sum_{v\in\mathcal{V}_1} \deg v + \sum_{v\in\mathcal{V}_2} \deg v.$$

Since each vertex in $\mathcal{V}_2$ has even degree, $\sum_{v\in\mathcal{V}_2} \deg v$ is even. This forces

$$\sum_{v\in\mathcal{V}_1} \deg v = 2|\mathcal{E}| - \sum_{v\in\mathcal{V}_2} \deg v$$

to be even. But for each $v \in \mathcal{V}_1$, the number $\deg v$ is odd, so there must be an even number of elements in $\mathcal{V}_1$, as otherwise the fact that the sum of an odd number of odd numbers is odd would cause a contradiction.

□

Definition 8.6. Suppose G is a graph with vertex set $\mathcal{V}$ and edge set $\mathcal{E}$, and G' is a graph with vertex set $\mathcal{V}'$ and edge set $\mathcal{E}'$. We say that G' is a **subgraph** of G if $\mathcal{V}' \subseteq \mathcal{V}$ and $\mathcal{E}' \subseteq \mathcal{E}$.

Definition 8.7. There are several concepts related to traveling through a graph:

- A **walk** on a graph is a list of vertices
$$(v_0, v_1, \ldots, v_k)$$
of the graph with $k \geq 1$, where it is guaranteed that $e_i = v_iv_{i+1}$ is always an edge of the graph for $0 \leq i \leq k-1$. We call it a walk between v_0 and v_k. The walk is **closed** if $v_0 = v_k$. Note that vertices are allowed to repeat in a walk.

- A **path** is a graph with vertex set $\mathcal{V} = \{v_0, v_1, \ldots, v_k\}$ and edge set
$$\mathcal{E} = \{v_0v_1, v_1v_2, \ldots, v_{k-1}v_k\}$$
for some integer $k \geq 1$. Note that the v_i must be distinct since they form a set. We denote the path by $v_0v_1 \cdots v_k$ and call it a path between v_0 and v_k. We will often talk about a path that is a subgraph of some graph, as opposed to a path by itself.

- A graph is called **connected** if, for every pair of distinct vertices, there is a subgraph of the graph that is a path between the vertices.

- A **cycle** is a graph with vertex set $\mathcal{V} = \{v_0, v_1, \ldots, v_k\}$ and edge set
$$\mathcal{E} = \{v_0v_1, v_1v_2, \ldots, v_{k-1}v_k, v_kv_0\}$$
for some integer $k \geq 2$. We denote the cycle by
$$v_0v_1 \cdots v_kv_0.$$
So a cycle is like a path, except it loops back to the first vertex. Like a path, a cycle does not travel though the same edge twice. We will often talk about a cycle that is a subgraph of some graph, as opposed to a standalone cycle.

- A connected graph that has no subgraphs that are cycles is called a **tree**.

Problem 8.8. A **circuit** is a walk that begins and ends on the same vertex and such that every edge is distinct. An **Eulerian circuit** is a circuit that visits every edge exactly once. Prove that, if an Eulerian circuit exists in a graph, then every vertex has even degree.

Theorem 8.9. There are several interesting properties of trees:

1. Every tree with at least two vertices has at least one leaf.

2. Every tree with $n \geq 1$ vertices has $n-1$ edges. As a result, every tree with $m \geq 0$ edges has $m+1$ vertices.

3. The sum of the degrees of the vertices of a tree with $n \geq 1$ vertices is $2n-2$.

4. Every tree with $n \geq 2$ vertices has at least two leaves, strengthening the first result.

Proof. We will prove the results in the given sequence, as later parts depend on earlier ones.

1. Suppose we have a tree with at least two vertices. Suppose, for contradiction, that it contains no leaf, meaning each vertex has at least two edges emanating from it. Among the finite number of all paths that are subgraphs of this tree, choose a maximal path, meaning the number of vertices in its vertex set is greater than or equal to the number of vertices in the vertex sets of all other paths. Picking either endpoint of this path, there must be an edge emanating from it that is not a part of the path because every vertex has degree at least 2. If the other endpoint of this edge is a vertex in the path, then this produces a cycle, which contradicts the fact that the graph is a tree. Otherwise, if that endpoint is not a vertex in the path, then it produces a longer path, which contradicts the maximality of our chosen path. Thus, a leaf must exist.

2. We will prove this by induction. In the base case $n = 1$, there can be no edge, as there is no second vertex. Now suppose the result holds for some integer $n \geq 1$. If we have a tree with $n+1$ vertices, we can remove a leaf (whose existence we established in the previous part) and the one edge emanating from it to produce a tree with n vertices. By the induction hypothesis, this new tree has $n-1$ edges and so our original tree has n edges.

3. Let the number of vertices in a tree be n, with the vertex set being $\mathcal{V}$ and the edge set being $\mathcal{E}$. By the handshaking lemma (Theorem 8.5),

$$\sum_{v \in \mathcal{V}} \deg v = 2|\mathcal{E}| = 2(n-1) = 2n-2.$$

4. Suppose, for contradiction, that there exists a tree with $n \geq 2$ vertices and there is exactly one leaf. So there are $n-1$ vertices of degree 2 or more and one vertex of degree 1, which, by the calculation in the previous part, leads us to the contradiction

$$2(n-1) = \sum_{v \in \mathcal{V}} \deg v \geq 2(n-1) + 1.$$

□

Lemma 8.10. There are two related lemmas that we will need with respect to traversing graphs:

1. If there is a walk between two distinct vertices, then there is a path between those two vertices.

2. Suppose we have a connected graph that contains a cycle as a subgraph. If an edge that belongs to the cycle is removed from the edge set of the graph, then the new graph remains connected.

Proof. We will prove the lemmas in sequence, with the first helping to prove the second.

1. Suppose there is a walk between two distinct vertices u and w. By the well-ordering principle, there is a walk between u and w with a minimum number of vertices among such walks. We claim that there are no repeated vertices in this walk. Let the sequence of vertices in this walk be

$$(v_0, v_1, \ldots, v_k)$$

where $u = v_0$ and $w = v_k$. If $k = 1$, then we are done, so suppose $k \geq 2$. For contradiction, suppose there is a repeated vertex, meaning for some indices i and j such that $0 \leq i < j \leq k$, we have $v_i = v_j$. But if we remove the subsequence $(v_{i+1}, v_{i+2}, \ldots, v_j)$ from $(v_0, v_1 \ldots, v_k)$, then we produce a walk between v_0 and v_k with strictly fewer vertices than our supposedly minimal walk, which is a contradiction. Thus, there are no repeated elements. So we can produce a path between v_0 and v_k with vertex set $\{v_0, v_1, \ldots, v_k\}$ and edge set $\{v_0v_1, v_1v_2, \ldots, v_{k-1}v_k\}$.

2. Let the cycle in question be $v_0v_1 \cdots v_kv_0$ and assume without loss of generality that the edge that is removed is v_kv_0, as otherwise we could simply cyclically relabel the vertices to force this. The main idea is that, even though v_0 and v_k are no longer connected by the path v_kv_0, we can go in the other direction of the cycle to connect them with the path $v_0v_1 \cdots v_k$. If two vertices in the original graph are connected by a path that does not include the edge v_kv_0, then this path still exists in the new graph. Now suppose u and w are vertices outside of the cycle that were connected by the path

$$u_0u_1 \cdots u_mv_0v_kw_0w_1 \cdots w_n$$

such that $u = u_0$ and $w = w_n$. Then we can connected them in the new graph with the walk

$$u_0u_1 \cdots u_mv_0v_1 \cdots v_kw_0w_1 \cdots w_n.$$

Then we have a path as well, since the existence of a walk implies the existence of a path, by the first part. The logic is similar if u or w or both are vertices on what used to be the cycle.

□

Problem 8.11. Prove the following results about connected graphs and trees using Lemma 8.10:

1. Every connected graph with $n \geq 1$ vertices has at least $n-1$ edges.
2. Every connected graph with exactly $n \geq 1$ vertices and exactly $n-1$ edges is a tree.

Theorem 8.12. Let $n \geq 2$ be an integer. By Theorem 8.9, the sum of the degrees of the vertices of a tree with n vertices is $2n-2$. Suppose $(d_1, d_2, \ldots, d_n)$ is a list of positive integers such that

$$\sum_{i=1}^{n} d_i = 2n-2.$$

Then the number of distinct trees with the vertex set $[n] = \{1, 2, \ldots, n\}$, such that each vertex i has degree d_i, is the multinomial coefficient

$$t(d_1, d_2, \ldots, d_n) = \binom{n-2}{d_1-1, d_2-1, \ldots, d_n-1}.$$

We consider two trees to be distinct if they have different edge sets.

Proof. We proceed by induction on $n \geq 2$. If $n = 2$, then d_1 and d_2 are positive integers such that $d_1+d_2 = 2$, so $d_1 = d_2 = 1$. There is only one connected graph on two vertices, which matches the computation

$$\binom{2-2}{1-1, 1-1} = \frac{0!}{0!0!} = 1.$$

For the induction hypothesis, assume the result holds for some integer $n \geq 2$. For the inductive step, suppose $(d_1, d_2, \ldots, d_{n+1})$ is a list of positive integers such that

$$\sum_{i=1}^{n+1} d_i = 2(n+1) - 2 = 2n.$$

If all of the d_i were greater than or equal to 2 then their sum would be greater than or equal to $2(n+1)$, which exceeds $2n$. So $d_k = 1$ for some $1 \leq k \leq n+1$. For convenience of notation, we will assume that $k = n+1$, and it will become evident that this is without loss of generality, as the argument can be adapted for $d_k = 1$ for any k, except with more cumbersome notation. Since $d_{n+1} = 1$, it is attached by an edge to exactly one other vertex $j \neq n+1$. Let $T(d_1, d_2, \ldots, d_{n+1})$ be the set of trees with vertex set $[n+1]$ and vertex i having degree d_i for each $1 \leq i \leq n+1$. For each $1 \leq j \leq n$, let $T_j(d_1, d_2, \ldots, d_n)$ be the subset of $T(d_1, d_2, \ldots, d_{n+1})$ such that vertex $n+1$ is attached only to vertex j. Then we have the disjoint union

$$T(d_1, d_2, \ldots, d_{n+1}) = \bigsqcup_{j=1}^{n} T_j(d_1, d_2, \ldots, d_n).$$

By removing vertex $n+1$ and its edge, we see that there is bijection between $T_j(d_1, d_2, \ldots, d_n)$ and trees with vertex set $[n]$ such that vertex i has degree d_i for $i \neq j$ and vertex j has degree $d_j - 1$. By the addition principle and the inductive hypothesis and the fact that $(d_{n+1}-1)! = (1-1)! = 0! = 1$,

$$\begin{aligned}
t(d_1, d_2, \ldots, d_{n+1}) &= |T(d_1, d_2, \ldots, d_{n+1})| \\
&= \sum_{j=1}^{n} |T_j(d_1, d_2, \ldots, d_n)| \\
&= \sum_{j=1}^{n} \frac{(n-2)!}{(d_1-1)! \cdots (d_j-2)! \cdots (d_n-1)!} \\
&= \sum_{j=1}^{n} \frac{(n-2)!(d_j-1)}{(d_1-1)! \cdots (d_j-1)! \cdots (d_n-1)!(d_{n+1}-1)!} \\
&= \frac{(n-2)!}{(d_1-1)! \cdots (d_{n+1}-1)!} \cdot \sum_{j=1}^{n} (d_j-1).
\end{aligned}$$

Since $d_{n+1} = 1$, we find that

$$\sum_{j=1}^{n}(d_j - 1) = \sum_{j=1}^{n+1}(d_j - 1) = \sum_{j=1}^{n+1} d_j - \sum_{j=1}^{n+1} 1 = 2n - (n+1) = n - 1,$$

and so

$$t(d_1, d_2, \ldots, d_{n+1}) = \frac{(n-2)!}{(d_1 - 1)! \cdots (d_{n+1} - 1)!} \cdot (n-1) = \frac{(n-1)!}{(d_1 - 1)! \cdots (d_{n+1} - 1)!},$$

which completes the induction. The algebraic manipulations at the end were essentially a repeat of those used to solve Problem 5.24. □

Problem 8.13 (Cayley's formula). Prove that, for $n \geq 1$, the number of distinct trees with the vertex set $[n] = \{1, 2, \ldots, n\}$ is n^{n-2}, where two trees are distinct if they have different edge sets.

8.2 Planar Graphs

Definition 8.14. A **planar graph** is a graph that can be drawn on the plane so that vertices are dots, edges are curves between vertices (where we do not include the endpoints as a part of an edge), and edges do not intersect each other. A particular way of drawing a planar graph on the plane in this way is called a **planar embedding** of the graph.

Some authors define planar graphs so that edges can only be *line segments* (minus endpoints), but Fáry's theorem states that this restriction does not result in any fewer graphs being admitted as planar. So, every planar graph has a planar embedding where the edges are line segments minus endpoints. From here on, we will always assume that edges in a planar embedding are line segments at the cost of weakening some results, in order to make their proofs easier to conceptualize.

Definition 8.15. Removing the vertices and edges of a planar embedding from the plane results in partitioning the remainder of the plane into regions that we call **faces** of the planar embedding. We can classify faces into two types, depending on where the face lies relative to the boundary of the graph, which consists of the outermost edges.

- There is one **unbounded** face that lies outside the boundary of the graph. Its boundary is the subgraph consisting of exactly the edges and vertices on the outer boundary of the graph.
- The remaining faces, of which there might be none, are **bounded** as they are enclosed by the outer boundary of the graph. The boundary of a bounded face is a subgraph consisting of the edges and vertices of the graph that are adjacent to the face.

Theorem 8.16 (Euler's formula for planar graphs). If V, E, F are the number of vertices, edges, and faces, respectively, of a planar embedding of a connected planar graph, then

$$\chi = V - E + F = 2.$$

Proof. We proceed by induction on the number of edges $\epsilon \geq 0$. The base case with 0 edges is easy because there is one vertex and the only face is the unbounded face, so

$$\chi = 1 - 0 + 1 = 2.$$

For the induction hypothesis, suppose the result holds for some number of edges $\epsilon \geq 0$. For the inductive step, suppose we have a planar embedding of a connected planar graph with $\epsilon + 1$ edges. We will split the inductive step into two cases: there is not a cycle in the graph and there is a cycle in the graph. If there is no cycle, then the graph is a connected graph with no cycles, which is a tree. We know that a tree with $\epsilon + 1$ edges has $\epsilon + 2$ vertices. Moreover, a planar embedding of a tree cannot have a bounded face because the boundary of a bounded face would produce a cycle in the graph; so the only face of a tree is the unbounded face. Thus, Euler's formula holds for trees, as

$$\chi = (\epsilon + 2) - (\epsilon + 1) + 1 = 2.$$

Now, suppose there is a cycle in the graph. Let V, E, F be the number of vertices, edges, and faces, respectively, in the given planar embedding of the graph. By Lemma 8.10, if an edge that belongs to this cycle is removed from the edge set of the graph, then the new graph remains connected; of course, removing an edge means the embedding of the new graph is still planar. The number of vertices V does not change and the number of edges decreases by 1 to $E-1$. Every edge in a cycle is shared by two faces, one on either side of the line running through the edge; removing the edge from the planar embedding results in these two faces becoming one face, so the number of faces decreases by 1 to $F-1$. This new graph has $E-1=\epsilon$ edges, so we can apply the induction hypothesis to it to get that

$$V-(E-1)+(F-1)=2,$$

which can be simplified to $\chi = V-E+F=2$. □

Definition 8.17. We can adapt the notion of the degree of a vertex to the degree of a face as follows. Suppose there is a planar embedding of a connected graph with at least two edges.

1. Each edge has a line running through it, and there is either a different face adjacent to the edge on either side of this line, or both sides of the line have the same face adjacent to the edge. So each edge is either adjacent to two faces, in which case we will call it **two-faced**, or adjacent to one face, in which case we will call it **one-faced**.

2. Contrary to intuition, we assign a value of 1 to each two-faced edge and a value of 2 to each one-faced edge. Then the **degree of a face** in a planar embedding is the sum of the values of the edges in the subgraph that is the boundary of the face. The degree of face f is denoted by $\deg f$. We chose this definition because it aligns with the number of edges between the vertices of a closed walk on the boundary of a face (try it out and see!); in particular, we need to traverse each one-faced edge twice and each two-faced edge once in such a walk.

Theorem 8.18 (Planarity necessary criterion)**.** Now we develop a necessary criterion for planarity. Suppose there is a connected, planar graph with at least two edges. Let V, E, F be the number of vertices, edges, and faces, respectively, in a given planar embedding of this graph. Then:

1. For each face f in the planar embedding, $\deg f \geq 3$.

2. If $\mathcal{F}$ is the set of faces in the given planar embedding, then an analogue of the handshaking lemma is that $\sum\limits_{f\in\mathcal{F}} \deg f = 2E$.

3. The planar embedding satisfies $2E \geq 3F$.

4. The graph satisfies $E \leq 3V-6$, independent of any particular planar embedding.

Proof. We assume the initial conditions in the theorem statement.

1. If f is a bounded face, then there are at least three edges on its boundary. Since the value of each edge is 1 or 2, we know that $\deg f \geq 3$. If f is the unbounded face, then we split the argument into the cases where the graph is a tree and where the graph contains a cycle. If the graph is a tree, then there is only one face, and so all the edges are one-faced and have a value of 2; since there are at least two edges, it implies $\deg f \geq 4$. If there is a cycle, then it is not possible for there to be only two edges on the outer boundary of the graph since two edges cannot enclose a region that contains a cycle. So there are at least three edges, from which the result follows.

2. Each edge is either two-faced or one-faced. Each two-faced edge contributes a value of 1 to the degrees of exactly two faces in the sum and nothing to the degrees of other faces. Each one-faced edge contributes a value of 2 to the degree of exactly one face and nothing to the degrees of other faces. Thus, the sum $\sum\limits_{f\in\mathcal{F}} \deg f$ counts each edge exactly twice, so it is equal to $2E$.

3. Using the last two parts,
$$2E = \sum_{f \in \mathcal{F}} \deg f \geq 3F.$$

4. By Euler's formula for planar graphs, $V - E + F = 2$ or $F = 2 - V + E$. Substituting this into $2E \geq 3F$ yields
$$2E \geq 3(2 - V + E)$$
or $E \leq 3V - 6$. The advantage of eliminating F like this is that this necessary criterion for planarity is independent of the number of faces in any particular embedding. The number of edges and the number of vertices does not depend on the planar embedding.

□

Example 8.19. The complete graph K_5 on 5 vertices is not planar.

Solution. There are $V = 5$ vertices and $E = \binom{5}{2} = 10$ edges in K_5. Then
$$E = 10 > 3 \cdot 5 - 6 = 3V - 6.$$

Since K_5 is connected, it cannot be planar by Theorem 8.18. □

Problem 8.20. Show that every connected planar graph has a vertex v such that $\deg v \leq 5$.

Definition 8.21. A graph is said to be **triangle-free** if any cycle in it contains 4 or more edges.

Problem 8.22 (Mantel's theorem). Let G be a triangle-free graph with vertex set $\mathcal{V}$ and edge set $\mathcal{E}$. Prove that
$$|\mathcal{E}| \leq \frac{|\mathcal{V}|^2}{4}.$$

Theorem 8.23. Suppose there is a connected, planar graph that is triangle-free and has at least two edges. Let V, E, F be the number of vertices, edges, and faces, respectively, in a given planar embedding of this graph. Then the graph satisfies $E \leq 2V - 4$, independent of any particular planar embedding. Note that this is stronger than Theorem 8.18 because
$$E \leq 2V - 4 \leq 3V - 6.$$

Of course, the trade-off of this strength is the requirement of being triangle-free.

Proof. Assume the initial conditions in the theorem. First, we will show that each face has degree greater than or equal to 4. If f is a bounded face, then its boundary contains a cycle as a subgraph, which we have assumed to have at least 4 edges (since it cannot be a triangle). So $\deg f \geq 4$. Now suppose f is the unbounded face. If the graph is a tree, we showed in the proof of Theorem 8.18 that $\deg f \geq 4$. If the graph is not a tree, then the graph contains a cycle. It is not possible for there to be only two or three edges edges on the outer boundary of the graph because this either means three edges form a cycle (that is, a triangle), or the two or three edges form a graph that cannot enclose a region that contains a cycle; either way, we have a contradiction. Thus, there must be at least four edges on the boundary of the unbounded face, which means $\deg f \geq 4$. If $\mathcal{F}$ is the set of faces of the graph, then the handshaking lemma for faces tells us that
$$2E = \sum_{f \in \mathcal{F}} \deg f \geq 4F$$
or $E \geq 2F$. By Euler's formula for planar graphs, $V - E + F = 2$ or $F = 2 - V + E$. Substituting this into $E \geq 2F$ yields
$$E \geq 2(2 - V + E)$$
or $E \leq 2V - 4$. □

Definition 8.24. A **bipartite** graph is a graph whose vertex set $\mathcal{V}$ may be partitioned into two sets $\mathcal{V}_1$ and $\mathcal{V}_2$ such that:

- $\mathcal{V}_1 \cap \mathcal{V}_2 = \varnothing$ and $\mathcal{V}_1 \cup \mathcal{V}_2 = \mathcal{V}$
- Every edge in the graph is between a vertex in $\mathcal{V}_1$ and a vertex in $\mathcal{V}_2$

Problem 8.25 (Three utilities problem). Prove the following:

1. Bipartite graphs do not contain cycles that have an odd number of edges. The converse, which says that if all cycles have an even number of edges then the graph is bipartite, also holds, but you are not asked to prove it.
2. Every bipartite graph is triangle-free.
3. Let the **Thomsen graph** $K_{3,3}$ be the graph with the vertex set $\{v_1, v_2, v_3, w_1, w_2, w_3\}$, and edge set $\{v_i w_j : 1 \leq i \leq 3, 1 \leq j \leq 3\}$. Then $K_{3,3}$ is not planar.

It is not a coincidence that we have chosen to disprove the existence of planar embeddings of K_5 (Example 8.19) and $K_{3,3}$ (Problem 8.25). This is related to a characterization of all planar graphs, called Kuratowski's theorem. The interested reader is encouraged to read about it elsewhere.

Definition 8.26. There are two concepts that we need, in order to apply Euler's formula to convex polyhedra:

- The **skeleton** of a convex polyhedron is the graph whose vertex set corresponds to the vertices of the polyhedron and whose edge set corresponds to the edges of the polyhedron. More precisely, if there is a skeleton edge between two skeleton vertices, then there is a polyhedron edge between the corresponding polyhedron vertices, and vice versa.
- A **Schlegel diagram** of a convex polyhedron is a planar embedding of its skeleton in such a way that the faces of the embedding correspond to the faces of the polyhedron. More precisely, if a face of the embedding is bounded by certain edges in the embedding, then the corresponding edges in the polyhedron also bound a face in the polyhedron, and vice versa.

Theorem 8.27. If the number of vertices, edges, and faces of a convex polyhedron are V, E, F, respectively, then $V - E + F = 2$.

Proof. It suffices to prove the existence of a Schlegel diagram for each convex polyhedron. Unfortunately, this is rather difficult to do rigorously, so we will not provide details. The idea amounts to using the convex polyhedron to produce the equivalent of a Schlegel diagram on a sphere, and then applying a transformation called "stereographic projection" to send the sphere to the plane, while preserving the vertex-edge-face structure at each step. Informally, one could say that we are holding a light source on top of the polyhedron's edges and vertices (with empty faces so that the light can go through) and observing the resulting shadow on the plane. □

Problem 8.28. In a given convex polyhedron, let V be the number of vertices and E be the number of edges. Prove that $2E \geq 3V$.

Definition 8.29. A **regular graph** is a graph whose vertices all have the same degree. If this common vertex degree is k, then we call it a k-regular graph.

Definition 8.30. A **regular polyhedron** or **Platonic solid** is a convex polyhedron whose faces are congruent regular polygons and each vertex has the same degree. Thus, the skeleton of a regular polyhedron with vertices of degree k is a k-regular graph.

Theorem 8.31. There are at most five ordered triples (V, E, F), representing the number of vertices V, edges E, and faces F of a regular polyhedron. These are recorded in the following table, where d_v is the degree of each vertex and d_f is the degree of each face.

Polyhedron	d_v	d_f	V	E	F
Tetrahedron	3	3	4	6	4
Hexahedron (Cube)	3	4	8	12	6
Octahedron	4	3	6	12	8
Dodecahedron	3	5	20	30	12
Icosahedron	5	3	12	30	20

Each of these five possibilities gives rise to a unique Platonic solid, but we will not show this.

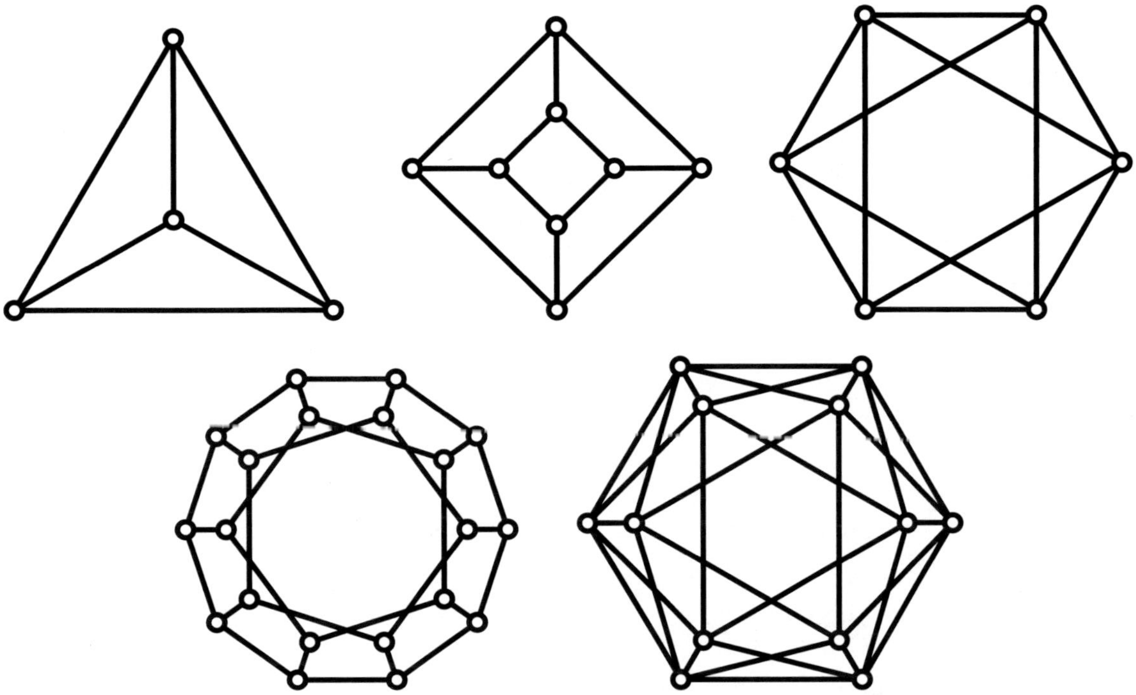

Proof. We present a proof that Coxeter included in [3]. Suppose we are given a regular polyhedron. Let $\mathcal{V}$ be the set of vertices and $\mathcal{F}$ be the set of faces. The handshaking lemma and its analogue for faces yields

$$Vd_v = \sum_{v\in\mathcal{V}} \deg v = 2E = \sum_{f\in\mathcal{F}} \deg f = Fd_f.$$

A ratio trick from Volume 1 tells us that

$$\frac{a}{b} = \frac{c}{d} \implies \frac{a}{b} = \frac{c}{d} = \frac{a+c}{b+d} = \frac{a-c}{b-d}$$

as long as we are never dividing by zero. Combined with Euler's formula (Theorem 8.16), this allows us to conclude

$$\frac{V}{\left(\frac{1}{d_v}\right)} = \frac{E}{\left(\frac{1}{2}\right)} = \frac{F}{\left(\frac{1}{d_f}\right)} = \frac{V-E+F}{\left(\frac{1}{d_v} - \frac{1}{2} + \frac{1}{d_f}\right)} = \frac{2}{\left(\frac{1}{d_v} - \frac{1}{2} + \frac{1}{d_f}\right)} = \frac{4d_vd_f}{4-(d_v-2)(d_f-2)}.$$

So we can find

$$V = \frac{4d_f}{4-(d_v-2)(d_f-2)},$$
$$E = \frac{2d_vd_f}{4-(d_v-2)(d_f-2)},$$
$$F = \frac{4d_v}{4-(d_v-2)(d_f-2)}$$

Now, we just need to find the possibilities for (d_v, d_f). These are severely restricted by the fact that they are each greater than or equal to 3 and must satisfy $(d_v - 2)(d_f - 2) < 4$ in order for the above expressions for V, E, F to be positive. So

$$1 \le (d_v - 2)(d_f - 2) \le 3,$$

which means we are looking for all solutions to the factored Diophantine equation $(d_v - 2)(d_f - 2) = k$ for $k = 1, 2, 3$. Based on the techniques of Volume 3, this yields the following solutions:

$$\begin{aligned}(d_v - 2)(d_f - 2) = 1 &\implies (d_v - 2, d_f - 2) = (1, 1),\\ (d_v - 2)(d_f - 2) = 2 &\implies (d_v - 2, d_f - 2) = (1, 2), (2, 1),\\ (d_v - 2)(d_f - 2) = 3 &\implies (d_v - 2, d_f - 2) = (1, 3), (3, 1).\end{aligned}$$

So we get the five solutions

$$(d_v, d_f) \in \{(3, 3), (3, 4), (4, 3), (3, 5), (5, 3)\}.$$

This overall deduction is usually done using an argument involving the interior angles of faces around a vertex, but we have provided an algebraic argument in the interest of reducing the necessary geometric prerequisites. Each pair (d_v, d_f) does lead to a triple of integers (V, E, F), as recorded in table in the statement of the theorem. □

Definition 8.32. There are different types of diagonals in a convex polyhedron:

- A **diagonal** is a line segment that is not an edge, but whose endpoints are vertices.
- A **face diagonal** is a diagonal that lies on the surface of the polyhedron.
- A **space diagonal** is a diagonal that is not a face diagonal, and so its interior lies in the interior of the polyhedron.

Example 8.33. Find the number of space diagonals of an icosahedron.

Solution. To find the total number of diagonals, we subtract the number of edges from the number of unordered pairs of vertices. This is $\binom{12}{2} - 30 = 36$. Since the faces are triangles, there are no face diagonals, so 36 is the number of space diagonals. □

Problem 8.34. Find the number of space diagonals of a dodecahedron.

8.3 Ramsey Theory

Ramsey theory is an area of math that is said to be about finding order within disorder. Perhaps less mysteriously, basic Ramsey theory can be interpreted as finding regularity within any edge colouring of a sufficiently large graph. The following is a "first" non-trivial Ramsey-type result.

Example 8.35 (Theorem on friends and strangers)**.** There are six people at a party. For each unordered pair of people, the pair is classified as "friends" if those two people have met previously or as "strangers" if they are meeting for the first time. So, each pair has a unique classification in this way. Show that there exist three people such that the three are pairwise friends or pairwise strangers. This is a special case of Ramsey's theorem (Theorem 8.40).

Solution. We represent the six people as the six vertices of a regular hexagon, though we do not connect any vertices (yet). For each unordered pair of people, if they are friends then we draw a blue line segment between the corresponding pair of vertices, and if they are strangers then we draw a red line segment between the corresponding pair of vertices. The goal is to show that there exists a triangle with three blue edges or three red edges.

We pick any vertex V. There exist five edges emanating from V. By the strong pigeonhole principle (Theorem 2.12), if we map the five edges to the two colours blue and red in any way, one of the two colours will receive at least $\left\lceil \frac{5}{2} \right\rceil = 3$ edges. Let that colour be α and let the other colour be β. Let the other endpoints of these three α-coloured edges emanating from V be A, B, C. If one of the edges AB, BC, CA, say UW, is α-coloured, then an α-coloured triangle UVW is produced. If none of the edges AB, BC, CA are α-coloured, then all of them are β-coloured, producing a β-coloured triangle ABC. Either way, there is a triangle whose edges all have the same colour. □

Now we will generalize the theorem on friends and strangers to Ramsey's theorem for 2 colours and furthermore to an arbitrary finite number of colours. There also exists an infinite variation of Ramsey's theorem that we will not touch upon.

Definition 8.36. Let n be a positive integer. A **complete graph** on n **vertices** is the graph with n vertices and an edge between every unordered pair of vertices; it is denoted by K_n. A **clique** in a graph is a subset of vertices such that every pair of distinct vertices in the subset are attached by an edge (so, every subset of a complete graph is a clique); a clique on m vertices is called an m**-clique**. An **edge colouring** of a complete graph assigns a colour to each edge from a given set of colours; formally, this can be thought to be a map from the edge set to a set of numbers which represent the colours. A **monochromatic** clique is a clique where the edges all have the same colour in the overarching assignment of colours to edges; if the colour is c, then the monochromatic clique may be called a c**-coloured** clique.

Definition 8.37. Let $c \geq 2$ be an integer, $[c] = \{1, 2, \ldots, c\}$ be c distinct "colours," and $(r_1, r_2, \ldots, r_c)$ be a list of integers greater than or equal to 2; formally, this is a function from $[c]$ to $\mathbb{Z}_{\geq 2}$. Suppose there exists a positive integer n such that in any edge colouring of K_n with the colours $[c]$, there exists an index $i \in [c]$ such that K_n contains a monochromatic clique on r_i vertices with all edges of colour i (and therefore this is true in any larger complete graph as well, since K_n is a subgraph of K_m if $n \leq m$). Given that such a positive integer n exists, the smallest such positive integer corresponding to c and $(r_1, r_2, \ldots, r_c)$ is called their **Ramsey number** and is denoted by $R(r_1, r_2, \ldots, r_c)$. We have avoided allowing any of the c_i to equal 1 because a 1-clique is a single vertex (so there are no edges), which is not a very meaningful concept for our purposes.

Problem 8.38. Show that, for any integers $r \geq 2$ and $b \geq 2$, $R(r, b)$ exists if and only if $R(b, r)$ exists, and that in this case, they are equal. Many sources vaguely cite "symmetry" but we are asking for an explicit argument here.

Problem 8.39. For each integer $t \geq 2$, prove that $R(2, t)$ and $R(t, 2)$ exist, and that both are equal to t. To clarify definitions, a 2-clique is a pair of vertices attached by an edge.

Our goal now will be to show that $R(r_1, r_2, \ldots, r_c)$ always exists by building recursive upper bounds.

Theorem 8.40 (Ramsey's theorem for two colours). For all integers $r \geq 2$ and $b \geq 2$, the Ramsey number $R(r, b)$ exists and, if $r \geq 3$ and $b \geq 3$, then it is bounded above by

$$R(r, b) \leq R(r-1, b) + R(r, b-1).$$

Proof. We will proceed by induction on $r + s \geq 2 + 2 = 4$. Problem 8.39 handles the case where $r + b = 4$ because it is equivalent to $r = b = 2$. For the induction hypothesis, suppose there exists an integer $m \geq 4$ such that $R(x, y)$ exists for all integers $x \geq 2$ and $y \geq 2$ satisfying $x + y = m$. Now suppose that the integers $r \geq 2$ and $b \geq 2$ satisfy $r + b = m + 1$. We want to prove that $R(r, b)$ exists. Since the cases where $r = 2$ or $b = 2$ are handled by Problem 8.39, we may assume that $r \geq 3$ and $b \geq 3$. Let the first colour be red and the second colour be blue, for convenience of communication. We claim that $R(r, b)$ exists and is bounded above by

$$n = R(r-1, b) + R(r, b-1),$$

where each summand exists by the induction hypothesis because

$$(r-1) + b = r + (b-1) = (r+b) - 1 = (m+1) - 1 = m.$$

In the complete graph K_n, let v be an arbitrary fixed vertex. We split the remaining vertices of K_n into two classes as follows. Let

$$\begin{aligned} P &= \text{ the set of vertices } w \text{ such that the edge } vw \text{ is red,} \\ Q &= \text{ the set of vertices } w \text{ such that the edge } vw \text{ is blue.} \end{aligned}$$

Then the set of vertices of K_n may be counted in two ways as the left and right sides of the identity

$$R(r-1,b) + R(r,b-1) = |P| + |Q| + 1,$$

where the $+1$ accounts for the vertex v. It must be true that $|P| \geq R(r-1,b)$ or $|Q| \geq R(r,b-1)$, for otherwise we would have both

$$\begin{aligned} |P| < R(r-1,b) &\implies |P| + 1 \leq R(r-1,b), \\ |Q| < R(r,b-1) &\implies |Q| + 1 \leq R(r,b-1), \end{aligned}$$

adding which would lead to the contradiction

$$|P| + |Q| + 1 < |P| + |Q| + 2 \leq R(r-1,b) + R(r,b-1).$$

Now we do casework on the two possibilities, along with two subcases in each:

1. Suppose $|P| \geq R(r-1,b)$. By the definition of this Ramsey number, P has a red $(r-1)$-clique or a blue b-clique.
   (a) If P contains a red $(r-1)$-clique, then $P \cup \{v\}$ contains a red r-clique, since v is attached to every $w \in P$ by a red edge.
   (b) If P contains a blue b-clique, then we are done.
2. Suppose $|Q| \geq R(r,b-1)$. By the definition of this Ramsey number, Q has a red r-clique or a blue $(b-1)$-clique.
   (a) If Q contains a red r-clique, then we are done.
   (b) If Q contains a blue $(b-1)$-clique, then $Q \cup \{v\}$ contains a blue b-clique, since v is attached to every $w \in Q$ by a blue edge.

Thus, a red r-clique or a blue b-clique exists in K_n, completing the inductive step and therefore finishes the proof. □

Corollary 8.41. For each pair of integers $r \geq 2$ and $b \geq 2$,

$$R(r,b) \leq \binom{r+b-2}{r-1}.$$

Proof. We proceed by induction on $r + b \geq 2$. If $r = 2$, then, by Problem 8.39,

$$R(2,b) = b = \binom{b}{b-1} = \binom{2+b-2}{b-1},$$

and a similar argument works for $b = 2$. These observations include the base case $r + b = 4$, since it is equivalent to $r = b = 2$. For the induction hypothesis, suppose there exists an integer $m \geq 4$ such that for all integers $x \geq 2$ and $y \geq 2$ satisfying $x + y = m$, it holds that

$$R(x,y) \leq \binom{x+y-2}{x-1}.$$

Now suppose $r \geq 2$ and $b \geq 2$ are integers such that $r + b = m + 1$. We have already taken care of the case where $r = 2$ or $b = 2$, so we may assume that $r \geq 3$ and $b \geq 3$. By the recursive inequality in Theorem 8.40,

$$\begin{aligned} R(r,b) &\leq R(r-1,b) + R(r,b-1) \\ &\leq \binom{(r-1)+b-2}{(r-1)-1} + \binom{r+(b-1)-2}{r-1} \\ &= \binom{r+b-3}{r-2} + \binom{r+b-3}{r-1} \\ &= \binom{r+b-2}{r-1}, \end{aligned}$$

where we used Pascal's identity (Corollary 4.6) in the final step. As a side note, for $r = b = n + 1$, the upper bound is the central binomial coefficient $\binom{2n}{n}$. □

Definition 8.42. Given a positive integer n, and a list

$$(a_1, a_2, \ldots, a_n),$$

we say that a **sublist** is a list

$$(a_{i_1}, a_{i_2}, \ldots, a_{i_m})$$

such that $i_1 < i_2 < \ldots < i_m$ (so it is necessary that $1 \leq m \leq n$). If the entries of the original list are real numbers, then this sublist is said to be

$$\begin{cases} \textbf{monotonically increasing} & \text{if } a_{i_1} < a_{i_2} < \cdots < a_{i_m} \\ \textbf{monotonically decreasing} & \text{if } a_{i_1} > a_{i_2} > \cdots > a_{i_m} \end{cases}.$$

In either case, the sublist is said be a **monotonic sublist**.

Corollary 8.43. Let n be a positive integer and let $(a_1, a_2, \ldots, a_n)$ be a list of n distinct real numbers. If $n \geq R(r,b)$ for some integers $r \geq 2$ and $b \geq 2$, then there exists a strictly increasing sublist of r elements or a strictly decreasing sublist of b elements.

Proof. Suppose n is a positive integer and $r \geq 2$ and $b > 2$ are integers such that $n \geq R(r,b)$. Let the first colour for $R(r,b)$ be red and the second colour be blue. Let $(a_1, a_2, \ldots, a_n)$ be as stated. The technique is to have n vertices denoted by the elements of $[n] = \{1, 2, \ldots, n\}$, and, for any two indices $i < j$ in $[n]$, colour the edge

$$ij \text{ with } \begin{cases} \text{red if} & a_i < a_j \\ \text{blue if} & a_i > a_j \end{cases}.$$

By this definition of the colouring of K_n, in any (not necessarily monochromatic) clique of this graph, we find a strict total order on the numbers a_i represented by the vertices, since every size comparison is encoded into the graph's colouring and a clique means the size comparison between the numbers represented by any two vertices is known. Specifically, by the definition of $R(r,b)$, there exists a red r-clique or a blue b-clique in this colouring. Now we do casework:

1. Suppose there exists a red r-clique. The strict total order on the corresponding a_i follows the rule that, if $i < j$, then $a_i < a_j$. So if the indices of the red r-clique are $i_1 < i_2 < \cdots < i_r$, then

$$a_{i_1} < a_{i_2} < \cdots < a_{i_r}.$$

2. Similarly, suppose there exists a blue b-clique. The strict total order on the corresponding a_i follows the rule that, if $i < j$, then $a_i > a_j$. So if the indices of the red r-clique are $j_1 < j_2 < \cdots < j_b$, then

$$a_{j_1} > a_{j_2} > \cdots > a_{j_b}.$$

So there exists a strictly increasing r-length sublist or a strictly decreasing s-length sublist. □

The preceding consequence (Corollary 8.43) of Ramsey's theorem for monotone sequences can be refined by a separate argument as follows, in the sense that the lower bound for the required length of the original sequence can be lowered.

Theorem 8.44 (Erdős-Szekeres theorem)**.** Let r and s be non-negative integers. Given a list of distinct real entries and length strictly greater than rs, there exists a monotonically increasing sublist of length strictly greater than r or a monotonically decreasing sublist of length strictly greater than s.

Proof. If $r = 0$ or $s = 0$, then the conclusion is easily attained by choosing any single-entry sublist, which is simultaneously monotonically increasing and monotonically decreasing because there is only one entry in the sublist. So, we may assume that r and s are positive integers now.
Let the original list be $(a_1, a_2, \ldots, a_n)$ where $n > rs$ for some non-negative integers r and s. For each a_k, we define the ordered pair (p_k, q_k), where p_k is the length of the longest increasing sublist whose first entry is a_k, and q_k is the length of the longest decreasing sublist whose first entry is a_k. Note that it is always true that $p_k \geq 1$ and $q_k \geq 1$ because a sublist is allowed to have only one entry. We claim that $f : a_k \mapsto (p_k, q_k)$ is injective. Suppose i and j are distinct indices such that $1 \leq i < j \leq n$. Since the entries of the original list are distinct, either $a_i < a_j$ or $a_i > a_j$. If $a_i < a_j$ then a_i can be tacked on to the beginning of the longest monotonically increasing sublist that starts with a_j to produce a longer monotonically increasing sublist that starts with a_i; so $p_i > p_j$. Similarly, if $a_i > a_j$ then a_i can be tacked on to the beginning of the longest monotonically decreasing sublist that starts with a_j to produce a longer monotonically decreasing sublist that starts with a_i; so $q_i > q_j$. Either way, $(p_i, q_i) \neq (p_j, q_j)$, proving the injectivity of f.
Now suppose, for contradiction, that all the p_k are integers from $[r]$ and that all the q_k are integers from $[s]$. Then there are rs possible pairs (p_k, q_k), the collection of which we can choose to be the codomain of f. Since f is injective with a domain of cardinality n and a codomain of cardinality rs, the pigeonhole principle says that $n \leq rs$. This contradicts the assumption that $n > rs$. Thus, there exists an index $1 \leq k \leq n$ such that $p_k > r$ or $q_k > s$.
For those who have noticed that r and s may be interchanged in the expression rs, it is an immediate extension of the theorem that there exists a monotonic sublist (either increasing or decreasing) of length greater than $\min\{r, s\}$. □

Problem 8.45. Prove that the Erdős-Szekeres theorem (Theorem 8.44) is stronger than Corollary 8.43. One reformulation of Erdős-Szekeres asserts that a list of $(r-1)(b-1)+1$ distinct real entries has a strictly increasing sublist of r entries or a strictly decreasing sublist of b entries. On the other hand, Corollary 8.43 makes the same conclusion, but with an overarching list of $R(r, b)$ entries. So, the reader is asked to prove that $(r-1)(b-1)+1 \leq R(r, b)$, or rather $(r-1)(b-1) < R(r, b)$.

Now we will extend Ramsey's theorem from 2 colours to an arbitrarily finite number of colours c.

Problem 8.46. As a generalization of Problem 8.38, show that, for any integers $c \geq 2$ and $r_1, r_2, \ldots, r_c \geq 2$ and any bijection $\sigma : [c] \to [c]$,

$$R(r_1, r_2, \ldots, r_c)$$

exists if and only if

$$R(r_{\sigma(1)}, r_{\sigma(2)}, \ldots, r_{\sigma(c)})$$

exists, and that in this case, they are equal.

Problem 8.47. Let $c \geq 2$ be an integer. For each integer $t \geq 2$, prove that $R(\underbrace{2, \ldots, 2}_{c-1 \text{ of 2's}}, t)$ exists and equals t. As in Problem 8.39, a 2-clique is a pair of vertices attached by an edge. Note that, by the general symmetry result (Problem 8.46), t can be swapped with any of the 2's without altering the result.

Theorem 8.48 (Ramsey's theorem for c colours)**.** For any integer $c \geq 2$ and integers $r_1, r_2, \ldots, r_c \geq 2$, the Ramsey number $R(r_1, r_2, \ldots, r_c)$ exists and, if $c \geq 3$, then it is bounded above by

$$R(r_1, r_2, \ldots, r_c) \leq R(r_1, r_2, \ldots, r_{c-2}, R(r_{c-1}, r_c))$$

Proof. We proceed by induction on the number of colours $c \geq 2$. The base case, where we have $c = 2$ colours, was treated in Theorem 8.40. For the induction hypothesis, suppose the result is true for some number of colours $d \geq 2$. Now we will prove the result for $c = d + 1$ number of colours. Let the colours correspond to the elements of $[c] = \{1, 2, \ldots, c\}$. We claim that, for any integers $r_1, r_2, \ldots, r_c \geq 2$, $R(r_1, r_2, \ldots, r_c)$ exists and is bounded above by

$$n = R(r_1, r_2, \ldots, r_{c-2}, R(r_{c-1}, r_c)).$$

Suppose K_n is an arbitrary complete graph whose edges are coloured by the colours in $[c]$. First we map this to another coloured K_n that is almost the same, except every edge coloured by c has its colour replaced by $c - 1$. So, there are only $c - 1$ usable colours now on the edges of the new graph, which will allow us to invoke the induction hypothesis. By the definition of n (which is a Ramsey number on $c - 1$ entries or colours), there exists an index $i \in [c - 2]$ such that there exists a monochromatic i-coloured r_i-clique or there exists a monochromatic $(c - 1)$-coloured $R(r_{c-1}, r_c)$-clique in the new graph. In the former case, we have the desired conclusion. In the latter case, we return to the original graph, where we now know there is an $R(r_{c-1}, r_c)$-clique whose edges are coloured only by $c - 1$ and c. The base case asserts that there exists a monochromatic $(c - 1)$-coloured r_{c-1}-clique or a monochromatic c-coloured r_c-clique, In either case, we are done. □

Corollary 8.49. Let $c \geq 2$ be an integer. Then for any integers $r_1, r_2, \ldots, r_c \geq 2$, it holds that

$$R(r_1, r_2, \ldots, r_c) \leq \binom{r_1 + r_2 + \cdots + r_c - c}{r_1 - 1, r_2 - 1, \ldots, r_c - 1}.$$

A variant of this result was first observed and communicated to the author by Kaixin Wang.

Proof. We proceed by induction on the number of colours $c \geq 2$. The base case, where $c = 2$, was already established in Corollary 8.41. Suppose the result holds for some integer $c \geq 2$. Now let $r_1, r_2, \ldots, r_{c+1} \geq 2$ be integers. As an edge case, if any of the r_i are equal to 2, then (by Problem 8.46, assume without loss of generality that $r_{c+1} = 2$), we can use the recursive bound in Theorem 8.48 and the computation in Problem 8.39 to get

$$R(r_1, r_2, \ldots, r_c, 2) \leq R(r_1, r_2, \ldots, R(r_c, 2)) = R(r_1, r_2, \ldots, r_c).$$

Then, by the induction hypothesis,

$$R(r_1, r_2, \ldots, r_c) \leq \binom{r_1 + r_2 + \cdots + r_c - c}{r_1 - 1, r_2 - 1, \ldots, r_c - 1}.$$

This upper bound may be weakened to the correct form

$$\binom{r_1 + r_2 + \cdots + r_c + 2 - (c + 1)}{r_1 - 1, r_2 - 1, \ldots, r_c - 1, 2 - 1}$$

by working backwards:

$$\begin{aligned} \binom{r_1 + r_2 + \cdots + r_c - c}{r_1 - 1, r_2 - 1, \ldots, r_c - 1} &\leq \binom{r_1 + r_2 + \cdots + r_c + 2 - (c + 1)}{r_1 - 1, r_2 - 1, \ldots, r_c - 1, 2 - 1} \\ &\iff 1 \leq r_1 + r_2 + \cdots + r_c + 2 - (c + 1) \\ &\iff c \leq r_1 + r_2 + \cdots + r_c. \end{aligned}$$

Within the induction on $c \geq 2$, we will perform induction on $r_1 + r_2 + \cdots + r_{c+1} \geq 2(c + 1)$. The above observations include the base case $r_1 + r_2 + \cdots + r_{c+1} = 2(c + 1)$, since it is equivalent to $r_i = 2$ for all $i \in [c + 1]$. For the induction hypothesis, suppose there exists an integer $m \geq 2(c + 1)$ such that for all integers $r_1, r_2, \ldots, r_{c+1} \geq 2$ satisfying $r_1 + r_2 + \cdots + r_{c+1} = m$, it holds that

$$R(r_1, r_2, \ldots, r_{c+1}) \leq \binom{r_1 + r_2 + \cdots + r_c - c}{r_1 - 1, r_2 - 1, \ldots, r_{c+1} - 1}.$$

Now suppose $r_1, r_2, \ldots, r_{c+1} \geq 2$ are integers such that $r_1 + r_2 + \cdots + r_{c+1} = m+1$. We have already taken care of the edge case where one or more of the r_i equals 2, so we may assume that $r_i \geq 3$ for each $i \in [c+1]$. A slight modification of the argument in Theorem 8.40 allows us to develop the recursive bound

$$R(r_1, r_2, \ldots, r_{c+1}) \leq R(r_1 - 1, r_2, \ldots, r_{c+1}) + R(r_1, r_2 - 1, \ldots, r_{c+1}) + \cdots + R(r_1, r_2, \ldots, r_{c+1} - 1).$$

Therefore, using the multinomial generalization of Pascal's identity (Problem 5.24), along with the induction hypothesis, completes the proof in a fashion similar to the ending of the proof of Corollary 8.41. The inner induction and the outer induction are completed simultaneously. $\square$

To illustrate the usefulness of Ramsey's theorem for c colours, we will show that Fermat's Last Theorem for any specific exponent is false modulo all sufficiently large primes. We will need to pull in some ideas from modular arithmetic, which will be covered in detail in several chapters of Volume 3. Just like in Corollary 8.43, we will need to produce specific clever colouring assignments for edges, tailored to a particular situation, whereas Ramsey's theorem works for all colouring assignments. In other words, the full power of Ramsey's theorem is not always necessary.

Corollary 8.50 (Schur's theorem in Ramsey theory)**.** For any integer $c \geq 2$, there exists an integer $n \geq 3$ such that, for any colouring of $[n] = \{1, 2, \ldots, n\}$ by c colours (not every colour needs to be used), there exist $x, y, z \in [n]$ of the same colour and $x + y = z$. (As a result, any larger n works too for this c.)

Proof. Let the c colours be represented by the numbers in $[c] = \{1, 2, \ldots, c\}$. The fact that we care only about the existence of a monochromatic triple, and not of any specific colour in $[c]$, makes us try for

$$n = R(\underbrace{3, 3, \ldots, 3}_{c \text{ entries of } 3}).$$

Let $\kappa : [n] \to [c]$ be an arbitrary colouring of $[n]$. The idea is to use the fact that, for any integers α, β, γ, it algebraically simplifies that

$$(\beta - \alpha) + (\gamma - \beta) = (\gamma - \alpha).$$

In K_n, colour each edge ij with $\kappa(|i - j|)$. By the definition of $n = R(\underbrace{3, 3, \ldots, 3}_{c \text{ entries of } 3})$, there exists a monochromatic triangle (three edges corresponding to three vertices) in this colouring of K_n. Let the vertices of this triangle be $1 \leq \alpha < \beta < \gamma \leq n$. Then

$$\kappa(\beta - \alpha) = \kappa(\gamma - \beta) = \kappa(\gamma - \alpha),$$
$$(\beta - \alpha) + (\gamma - \beta) = (\gamma - \alpha).$$

Since α, β, γ are distinct integers in $[n]$, it is also true that

$$x = \beta - \alpha, y = \gamma - \beta, z = \gamma - \alpha$$

are integers in $[n]$. So the integers x, y, z chosen here satisfy the stated conditions. $\square$

Corollary 8.51 (Fermat's last theorem modulo p)**.** For every positive integer n, there exists a positive integer m such that, for all primes $p \geq m$, there exist integers x, y, z such that

$$x^n + y^n \equiv z^n \pmod{p}$$

and none of x, y, z are divisible by p. We exclude the possibility of solutions where at least one of x, y, z is divisible by p because then that number reduces to 0 modulo p, which produces trivial solutions.

Proof. There seems to be a connection to Schur's theorem in Ramsey theory (Corollary 8.50) because two numbers are being added to produce a third, albeit modulo p. Moreover, the section

$$[p - 1] = \{1, 2, \ldots, p - 1\}$$

of the positive integers forms a reduced residue system modulo p (this is partially why we require primes instead of arbitrary moduli in this proof). It from from this set $[p-1]$ that we will be picking x, y, z modulo p. Since p is a prime, there exists a multiplicative generator g of $[p-1]$ modulo p (this is the other reason for why it is convenient for the modulus to be prime; see Volme 3 for material on primitive roots). This means that, for each $w \in [p-1]$, there exists a unique integer exponent $t_w \in [p]$ such that

$$w \equiv g^{t_w} \pmod{p}.$$

To get an exponent of n on g, we perform the Euclidean division of t_w by n to get the unique quotient q_w and the unique remainder r_w such that

$$t_w = nq_w + r_w, \text{ and } 0 \leq r_w < n.$$

If we could find three elements where two add up to each other and the remainders of their exponents, as defined above, are all equal to r, then we would be done because we could cancel out the g^r (which is invertible modulo p) and get a solution to the equation.
Our technique of getting the exponents to equal to each other will be to link them to colouring, and to get a monochromatic triple of integers via Schur's theorem. Let there be n colours corresponding to the integers in $[n]$. We colour each element $w \in [p-1]$ so that if $w \equiv g^{nq_w+r_w} \pmod{p}$, then w is coloured with the distinct colour $r_w + 1 \in [n]$. Note that the $+1$ is to account for the fact that

$$0 \leq r_w < n \iff 0 \leq r_w \leq n-1 \iff r_w + 1 \in [n].$$

By Schur's theorem (Corollary 8.50), there exist $\alpha, \beta, \gamma \in [p-1]$ such that they are monochromatic and $\alpha + \beta = \gamma$. By the fact that they have the same colour, we get

$$r_\alpha = r_\beta = r_\gamma,$$

so let the common value of the remainders be r. Then

$$\begin{aligned} \alpha + \beta &= \gamma \\ \alpha + \beta &\equiv \gamma \pmod{p} \\ g^{t_\alpha} + g^{t_\beta} &\equiv g^{t_\gamma} \pmod{p} \\ g^{nq_\alpha + r} + g^{nq_\beta + r} &\equiv g^{nq_\gamma + r} \pmod{p} \\ (g^{q_\alpha})^n + (g^{q_\beta})^n &\equiv (g^{q_\gamma})^n \pmod{p}. \end{aligned}$$

So we can choose x, y, z to be the reduced versions of $g^{q_\alpha}, g^{q_\beta}, g^{q_\gamma}$ modulo p, respectively, though reduction is not strictly necessary, as any integers will do. ☐

In Corollary 8.50, the quantity $R_c(3) = R(\underbrace{3, 3, \ldots, 3}_{c \text{ entries of } 3})$ came up. It is an interesting number, as it represents the minimum number of vertices at which we are guaranteed to have a monochromatic triangle, without caring about the specific colour of the triangle. We will find an upper bound on this number now.

Lemma 8.52. Let $c \geq 2$ be an integer, and let $f(c)$ be the maximum number such that the edges of the complete graph on $f(c)$ vertices may be coloured using c colours in a way that it does not contain a monochromatic triangle. Then

$$f(c+1) \leq (c+1) \cdot f(c) + 1.$$

Note that $R_c(3) = f(c) + 1$.

Proof. Suppose, for contradiction, that

$$f(c+1) > (c+1) \cdot f(c) + 1.$$

Given a complete graph on $f(c+1)$ vertices with an arbitrary edge colouring, we want to show that there necessarily exists a monochromatic triangle under the aforementioned assumption. Fixing any vertex v, there exist $f(c+1) - 1$ vertices edges emanating from v. By assumption,

$$f(c+1) - 1 \geq (c+1) \cdot f(c) + 1,$$

so v is connected to at least $(c+1)\cdot f(c)+1$ other vertices. As there are $c+1$ colours, the strong pigeonhole principle (Theorem 2.12) tells us that there exist at least

$$\left\lceil \frac{(c+1)f(c)+1}{c+1} \right\rceil = f(c)+1$$

edges emanating from v that have the same colour b. If two of those vertices are connected by an edge of colour b, then then we find a monochromatic triangle. Otherwise, all of the edges between those $f(c)+1$ vertices are coloured by only c colours (since one colour is assumed to be missing). As $f(c)+1 > f(c)$, the definition of f dictates that a monochromatic triangle exists among the edges between these vertices. $\square$

Lemma 8.53. If $c \geq 2$ is an integer, then

$$\{e \cdot c!\} = c! \cdot \sum_{i=c+1}^{\infty} \frac{1}{i!},$$

where the curly brackets denote the fractional part of the enclosed expression.

Proof. The infinite series definition of e is

$$e = 1 + \frac{1}{1!} + \frac{1}{2!} + \frac{1}{3!} + \frac{1}{4!} + \cdots.$$

Then

$$e \cdot c! = \left(c! + \frac{c!}{1!} + \frac{c!}{2!} + \frac{c!}{3!} + \cdots + \frac{c!}{c!} \right) + \frac{c!}{(c+1)!} + \frac{c!}{(c+2)!} + \cdots,$$

where the initial terms inside the brackets are all integers. We will now prove that the fractional part of $e \cdot c!$ is the sum of the remaining terms by showing that those terms add up to less than 1. Indeed,

$$\begin{aligned}
\frac{c!}{(c+1)!} + \frac{c!}{(c+2)!} + \frac{c!}{(c+3)!} + \cdots &= \frac{1}{c+1} + \frac{1}{(c+1)(c+2)} + \frac{1}{(c+1)(c+2)(c+3)} + \cdots \\
&\leq \frac{1}{c+1} + \frac{1}{(c+1)(c+2)} + \frac{1}{(c+2)(c+3)} + \frac{1}{(c+3)(c+4)} \cdots \\
&= \frac{1}{c+1} + \sum_{i=c+1}^{\infty} \frac{1}{i(i+1)} \\
&= \frac{1}{c+1} + \sum_{i=c+1}^{\infty} \left(\frac{1}{i} - \frac{1}{i+1} \right) \\
&= \frac{1}{c+1} + \frac{1}{c+1} = \frac{2}{c+1} < 1,
\end{aligned}$$

where we used $c \geq 2$ in the last step and telescoping via partial fraction decomposition before that. $\square$

Theorem 8.54. If $c \geq 2$ is an integer, then

$$R_c(3) = R(\underbrace{3, 3, \ldots, 3}_{c \text{ entries of } 3}) \leq \lfloor e \cdot c! \rfloor + 1.$$

Proof. We will proceed by induction on the number of colours $c \geq 2$. We note that, although taking off the floor function would produce a bound of equal strength (since the left side of the inequality is a Ramsey number, which is an integer), the inductive step would not work out on this case. The reader is encouraged to try it out. Instead, we will work with the floor function as given. The base case states that

$$R_2(3) \leq \lfloor e \cdot 2! \rfloor + 1 \leq 6,$$

which is true due to the theorem on friends and strangers (Example 8.35). For the induction hypothesis, suppose there exists an integer $c \geq 2$ such that $R_c(3) \leq \lfloor e \cdot c! \rfloor + 1$. In the notation of Lemma 8.52, this is equivalent to $f(c) \leq \lfloor e \cdot c! \rfloor$. We wish to prove that $R_{c+1}(3) \leq \lfloor e \cdot (c+1)! \rfloor + 1$, which is equivalent to

$f(c+1) \leq \lfloor e \cdot (c+1)! \rfloor$, in the notation of the lemma. Multiplying both sides of the induction hypothesis by $c+1$ and adding 1, we get

$$(c+1)f(c) + 1 \leq (c+1)\lfloor e \cdot c! \rfloor + 1.$$

By the lemma, $f(c+1) \leq (c+1)f(c) + 1$, so it suffices to prove that

$$(c+1)\lfloor e \cdot c! \rfloor + 1 \leq \lfloor e \cdot (c+1)! \rfloor.$$

By Lemma 8.53, $\{e \cdot c!\} > \dfrac{1}{c+1}$, since the right side is the first term in this lemma. Then

$$\lfloor e \cdot c! \rfloor = e \cdot c! - \{e \cdot c!\} < e \cdot c! - \frac{1}{c+1} = \frac{e \cdot (c+1)! - 1}{c+1} < \frac{\lfloor e \cdot (c+1)! \rfloor}{c+1},$$

yielding

$$(c+1)\lfloor e \cdot c! \rfloor < \lfloor e \cdot (c+1)! \rfloor,$$

which is equivalent to the desired

$$(c+1)\lfloor e \cdot c! \rfloor + 1 \leq \lfloor e \cdot (c+1)! \rfloor.$$

□

The following problem addresses an aspect of the vertex colouring analogue of edge colouring.

Problem 8.55 (Chromatic polynomial). Given a graph G and $k \geq 0$ colours, let the **chromatic function** $\chi_G(k)$ of G denote the number of proper colourings of the vertices of G with k fixed colours, where "proper" means that no two vertices connected by an edge are coloured by the same colour. Note that it is not necessary that all k colours are used. Prove the following facts:

1. Let uv be an edge of a graph G. Let $G \backslash uv$ denote G with the edge uv removed but without any changes to the vertices. Let G/uv denote G with the vertices u and v "merged" so that there is a single vertex w in place of u, v and the set of neighbours of w is the union of the set of neighbours of u and the set of neighbours of v, with all other vertices and edges remaining the same. Prove that

$$\chi_G(k) = \chi_{G\backslash uv}(k) - \chi_{G/uv}(k).$$

   This is a variation of the deletion-contraction recurrence.

2. Prove that the number of proper colourings with k colours of a graph with n vertices and no edges is k^n. Use this with the deletion recurrence to prove by strong induction that the chromatic function $\chi_G(k)$ of any fixed graph G is a polynomial whose variable is the number of colours k. Hence, the chromatic function of G is frequently called the **chromatic polynomial** of G.

3. Prove that the chromatic polynomials of the follow graphs are as stated:

   (a) For the complete graph K_n with n vertices,

$$\chi_{K_n}(k) = k(k-1)(k-2)\cdots(k-(n-1)).$$

   (b) For a tree T with n vertices

$$\chi_T(k) = k(k-1)^{n-1}.$$

   Note that paths are trees, so the case of the path P_n with n vertices is included in this formula.

   (c) For the cycle C_n with n vertices,

$$\chi_{C_n}(k) = (k-1)^n + (-1)^n(k-1).$$

   As a hint, a method of computing χ_{C_n} that does not assume knowledge of the conclusion, and therefore is non-mechanical, utilizes the principle of inclusion-exclusion.

Chapter 9

Probability

"One has in practical life to act upon probabilities, and what I should look to philosophy to do is to encourage people to act with vigor without complete certainty."

– *Bertrand Russell*

"One day, when I was doing well in class and had finished my lessons, I was sitting there trying to analyze the game of tic-tac-toe... The teacher came along and snatched my papers on which I had been doodling... She did not realize that analyzing tic-tac-toe can lead into dozens of non-trivial mathematical questions."

– *Martin Gardner*

At an intuitive level, probability assigns a numerical value to an event so that the value corresponds to our feeling for the "chances" of the event taking place within some larger context. This has to be made precise and the ramifications of the definition need to be studied. We will be working with finite probability spaces, but there exists a grander theory which contains our exposition as a small, albeit important, case. After establishing the foundations of finite probability spaces, we will investigate two special topics within probability. The first is conditional probability, where we can refine our quantification of the probability of events if there is additional information available, specifically if some event is known to occur. Secondly, we will look at the concept of random variables, which associates real values, usually meaningful ones, to the elements of the sample space, and the expected value of a random variable, which determines an average of the values of a random variable weighed according to probability.

9.1 Probability Spaces

Definition 9.1. A **finite probability space** consists of a non-empty finite set Ω, called the **sample space**, that is imbued with a function $p : \Omega \to \mathbb{R}_{\geq 0}$, called the **atomic probability distribution**, such that

$$\sum_{\omega \in \Omega} p(\omega) = 1.$$

The elements $\omega \in \Omega$ are called **atomic events**. The power set $\mathcal{P}(\Omega)$ is called the **event space** with its elements (the subsets of Ω) being called **events**. Using p, we define another function, called the **probability distribution**,

$$\begin{aligned} \mathbb{P} : \mathcal{P}(\Omega) &\to \mathbb{R}_{\geq 0} \\ A &\mapsto \sum_{\omega \in A} p(\omega). \end{aligned}$$

For any event $A \subseteq \Omega$, the value $\mathbb{P}(A)$ is called the **probability** of A.
We have preferred the term "finite" to "discrete" because it is not unreasonable to apply the latter to cases where the sample space is countably infinite instead of finite. On a separate note, some authors require that all non-empty events have non-zero probability, but we have chosen otherwise because this way is more general.

Example. For any atomic event $\omega \in \Omega$, the probability of the event that is the singleton containing it is

$$\mathbb{P}(\{\omega\}) = p(\omega).$$

As a blending of the two expressions, it is acceptable to write $\mathbb{P}(\omega)$ in the place of $\mathbb{P}(\{\omega\})$.

For the entirety of the chapter, unless it is otherwise specified, we let the event space be Ω, the atomic probability distribution be p, and the probability distribution be $\mathbb{P}$. This way, we will not have to repeatedly define the setup.
In general probability spaces, it turns out that it is not always possible to take the entire power set of the sample space as the event space without running into issues. In finite probability spaces, since we can and do take the event space to be the power set of the sample space, it is easy to show that events have all of the expected closure properties over set operations: the union, intersection, difference, and complement of events are events, and any subset of an event is an event. As such, in contrast with how we were initially careful to prove that a set is finite before speaking of its cardinality, we will rarely prove that a set is an event before speaking of its probability.

Definition 9.2. The **uniform probability** is the probability distribution arising from defining the atomic probability distribution as $p(\omega) = \dfrac{1}{|\Omega|}$ for all $\omega \in \Omega$. Indeed, this satisfies

$$\sum_{\omega\in\Omega} p(\omega) = \sum_{\omega\in\Omega} \frac{1}{|\Omega|} = |\Omega| \cdot \frac{1}{|\Omega|} = 1.$$

Theorem 9.3. Having the uniform probability implies that, for any event A,

$$\mathbb{P}(A) = \sum_{\omega\in A} p(\omega) = \sum_{\omega\in A} \frac{1}{|\Omega|} = \frac{1}{|\Omega|} \cdot \sum_{\omega\in A} 1 = \frac{|A|}{|\Omega|},$$

which is why many probability problems boil down to two combinatorial problems: finding the cardinality of the event and the cardinality of the sample space. Unless otherwise specified, the probability distribution of a finite probability space is uniform. However, do not confuse *possibility* with *probability*, as not all finite probability distributions are uniform.

Example. In the case of uniform probability, a common technique which we would like to point out is that, if Ω is a symmetrically dependent set of n-tuples for some positive integer n with dependence numbers $d_1, d_2, \dots, d_n$, and A is a symmetrically dependent subset of Ω with dependence numbers $a_1, a_2, \dots, a_n$, then

$$\mathbb{P}(A) = \frac{|A|}{|\Omega|} = \frac{a_1 \cdot a_2 \cdots a_n}{d_1 \cdot d_2 \cdots d_n} = \frac{a_1}{d_1} \cdot \frac{a_2}{d_2} \cdots \frac{a_n}{d_n}.$$

This can be seen from a more sophisticated perspective in light of the chain rule for events (Theorem 9.18).

Theorem 9.4 (Kolmogorov axioms)**.** Finite probability spaces satisfy the three Kolmogorov axioms:

1. The probability of every event $A \in \mathcal{P}(\Omega)$ is bounded below as $\mathbb{P}(A) \geq 0$.

2. The probability of the sample space is $\mathbb{P}(\Omega) = 1$

3. For any positive integer n, if $(A_1, A_2, \dots, A_n)$ is an n-tuple of pairwise disjoint events (this means that, if i, j are distinct indices, then $A_i \cap A_j = \varnothing$), then

$$\mathbb{P}\left(\bigcup_{i=1}^{n} A_i\right) = \sum_{i=1}^{n} \mathbb{P}(A_i).$$

In fact, we can simplify this to stating only

$$\mathbb{P}(A \cup B) = \mathbb{P}(A) + \mathbb{P}(B)$$

if A, B are disjoint events because induction allows us to conclude the initially stated version for n events.

For general probability spaces, the third Kolmogorov axiom actually involves an infinite series, but what we have stated is the equivalent requirement for finite probability spaces.

Proof. The requirements are called axioms, but they are not assumptions for us. We can prove them from our definition of a finite probability space.

1. Since a codomain of p is $\mathbb{R}_{\geq 0}$, we get the lower bound
$$\mathbb{P}(A) = \sum_{\omega \in A} p(\omega) \geq \sum_{\omega \in A} 0 = 0.$$

2. By the definition of the atomic probability distribution, the probability of the sample space is
$$\mathbb{P}(\Omega) = \sum_{\omega \in \Omega} p(\omega) = 1.$$

3. We will prove this property by induction on $n \geq 1$. In the base case $n = 1$, it is clearly true that $\mathbb{P}(A_1) = \mathbb{P}(A_1)$. Suppose the result is true for some integer $n \geq 1$ and let
$$(A_1, A_2, \ldots, A_n, A_{n+1})$$
be an $(n+1)$-tuple of pairwise disjoint events. Using an argument from our proof of the addition principle (Theorem 1.31),
$$B = A_1 \cup A_2 \cup \cdots \cup A_n \text{ and } A_{n+1}$$
are disjoint. Then
$$\mathbb{P}(B \cup A_{n+1}) = \sum_{\omega \in B \cup A_{n+1}} p(\omega) = \sum_{\omega \in B} p(\omega) + \sum_{\omega \in A_{n+1}} p(\omega) = \mathbb{P}(B) + \mathbb{P}(A_{n+1}).$$
By the induction hypothesis,
$$\mathbb{P}(A_1 \cup A_2 \cup \cdots \cup A_n \cup A_{n+1}) = \mathbb{P}(B \cup A_{n+1}) = \mathbb{P}(B) + \mathbb{P}(A_{n+1}) = \sum_{i=1}^{n} \mathbb{P}(A_i) + \mathbb{P}(A_{n+1}),$$
where the inductive hypothesis was used in the computation of $\mathbb{P}(B)$.

For those wondering why we used a tuple in the third Kolmogorov axiom instead of a multiset like in the addition principle of combinatorics, it is because the general form of the third axiom says that if $(A_1, A_2, A_3, \ldots)$ is a sequence of pairwise disjoint events, then
$$\mathbb{P}\left(\bigcup_{i=1}^{\infty} A_i\right) = \sum_{i=1}^{\infty} \mathbb{P}(A_i).$$
While it turns out that every ordering of an infinite series leads to the same sum when all of the terms are non-negative (like for probabilities), it is necessary to set down a particular order in order to evaluate an infinite series according to the definition of convergence of a series. To be in compliance with the general form of the third Kolmogorov axiom, we have placed the A_i in a tuple instead of a multiset. □

Theorem 9.5. The following implications of Kolmogorov's axioms hold for finite probability spaces. Let A and B be events.

1. If $A \subseteq B$, then $\mathbb{P}(A) \leq \mathbb{P}(B)$. This property is called **monotonicity**.
2. The probability of the empty event is $\mathbb{P}(\varnothing) = 0$.
3. Just like complementary counting in combinatorics, there is **complementary probability**, which says that the probability of $\overline{A} = \Omega \backslash A$ is $\mathbb{P}(\overline{A}) = 1 - \mathbb{P}(A)$.
4. The upper bound $\mathbb{P}(A) \leq 1$ holds.

5. As a probabilistic analogue of the principle of inclusion exclusion for two sets (Theorem 1.37),
$$\mathbb{P}(A \cup B) = \mathbb{P}(A) + \mathbb{P}(B) - \mathbb{P}(A \cap B).$$

Proof. Let A and B be events. The reader may wish to review our work on addition and subtraction (Chapter 1) in combinatorics, as several of the results or techniques used at the time are relevant to these proofs. We will quote them as needed.

1. Suppose $A \subseteq B$. Based on our work on Venn diagrams (Lemma 1.36), we know that $B \backslash A$ and A are disjoint events whose union is $A \cup B$. By the first and third Kolmogorov axioms,
$$\mathbb{P}(A) \leq \mathbb{P}(A) + \mathbb{P}(B \backslash A) = \mathbb{P}(A \cup (B \backslash A)) = \mathbb{P}(A \cup B),$$
which is equal to $\mathbb{P}(B)$ because $A \subseteq B$ implies that $A \cup B = B$.

2. By the third Kolmogorov axiom,
$$\mathbb{P}(\varnothing) = \mathbb{P}(\varnothing \cup \varnothing) = \mathbb{P}(\varnothing) + \mathbb{P}(\varnothing).$$
We cancel $\mathbb{P}(\varnothing)$ from both sides to get $\mathbb{P}(\varnothing) = 0$. One might recognize the similarity with the proof of $a \cdot 0 = 0$ in a field (see Volume 1).

3. As stated when studying complementary counting (Theorem 1.34), A and $\overline{A}$ are disjoint sets whose union is Ω. By the second and third Kolmogorov axioms,
$$1 = \mathbb{P}(\Omega) = \mathbb{P}(A \cup \overline{A}) = \mathbb{P}(A) + \mathbb{P}(\overline{A}),$$
which is a rearrangement of the desired equation.

4. Since $\mathbb{P}(\overline{A}) \geq 0$ due to the the first Kolmogorov axiom, complementary probability tells us that
$$\mathbb{P}(A) = 1 - \mathbb{P}(\overline{A}) \leq 1.$$

5. The proof is almost identical to the proof of the principle of inclusion-exclusion for two sets. Recall from our study of Venn diagrams (Lemma 1.36) that
$$\begin{aligned} &(A \backslash B, B), \\ &(B \backslash A, A), \\ &(A \backslash B, A \cap B, B \backslash A) \end{aligned}$$
are each tuples of pairwise disjoint events such that the union of the entries of each tuple is $A \cup B$. By applying the third Kolmogorov axiom to each of the tuples, we get the equations
$$\begin{aligned} \mathbb{P}(A \backslash B) + \mathbb{P}(B) &= \mathbb{P}(A \cup B), \\ \mathbb{P}(B \backslash A) + \mathbb{P}(A) &= \mathbb{P}(A \cup B), \\ \mathbb{P}(A \cup B) &= \mathbb{P}(A \backslash B) + \mathbb{P}(A \cap B) + \mathbb{P}(B \backslash A). \end{aligned}$$
By adding the three equations and cancelling the terms common to both sides of the resulting equation, we get
$$\mathbb{P}(A) + \mathbb{P}(B) = \mathbb{P}(A \cup B) + \mathbb{P}(A \cap B),$$
which is equivalent to the desired equation.

□

We will use the following lemma to solve Bertrand's ballot problem (Theorem 9.7). The proof is a bit technical, but we feel that it is a valuable contribution to the text because we have seen the lemma crop up in several contexts where it is taken for granted without articulation.

Lemma 9.6 (Discrete intermediate value theorem)**.** Let $n \geq 3$ be an integer, and $f : [n] \to \mathbb{Z}$ be an n-tuple such that for every index k satisfying $1 \leq k \leq n-1$, it holds that

$$|f(i+1) - f(i)| \leq 1.$$

For any integer a and any two indices i, j such that $1 \leq i < j \leq n$, if $f(i) < a < f(i+1)$ then there exists an index k such that $i < k < j$ and $f(k) = a$.

Proof. We are inspired to look at the indices at which f takes on values *greater* than a and investigate what happens one step below the least such index. Formally, let

$$S = \{p \in \mathbb{Z}_+ \cap [i,j] : f(p) > a\},$$

which is non-empty because $f(j) > a$ and so $j \in S$. By the well-ordering principle, let m be the least element of S. We claim that $k = m - 1$ satisfies the two conditions: $i < k < j$ and $f(k) = a$.
First we will show that k is strictly between i and j. Since $m \in S$, $i \leq m \leq j$. If $m = i$, then

$$a < f(m) = f(i) < a,$$

which is a contradiction. So $i < m \leq j$, from which we get $i \leq m - 1 < m \leq j$ or

$$i \leq k < k+1 \leq j.$$

Suppose for contradiction that $k = i$. Then

$$f(k) = f(i) < a < f(m) = f(k+1),$$

leading to

$$f(k) + 1 \leq a \leq f(k+1) - 1$$

or $2 \leq f(k+1) - f(k)$. But we know that $i \leq k < k+1 \leq j$, so k and $k+1$ are both indices in $[n]$, leading to them satisfying

$$|f(k+1) - f(k)| \leq 1,$$

which is a contradiction. So, the assumption that $k = i$ is incorrect and we instead have $i < k$. Combining this with our earlier finding, we have proven $i < k < j$.
Finally, we need to prove that $f(k) = a$. Suppose for contradiction that $f(k) \neq a$. Then either $f(k) > a$ or $f(k) < a$. If $f(k) > a$, then k would contradict the minimality of $m = k+1$ as the least element of S. So, under the assumption that $f(k) \neq a$, it must be true that $f(k) < a$. But then we get

$$f(k) < a < f(m) = f(k+1)$$

again, which we have shown to lead to a contradiction earlier in the argument. So $f(k) = a$, and we are done. □

Readers familiar with calculus might recognize the similarity of Lemma 9.6 with the well-known intermediate value theorem from calculus, but the latter applies only to continuous functions.

Theorem 9.7 (Bertrand's ballot problem)**.** There is an election between two candidates, A and B. After the ballots have been counted, it turns out that there are a votes for A and b votes for B such that $a > b$, where a and b are non-negative integers that add up to a positive integer (for those wondering, the vote is binary, so there are no abstained votes or spoiled ballots). When counting the ballots, the counters have to go through all $a+b$ ballots in some sequence. Then, out of all possible $(a+b)$-tuples that contain a ballots in favour of A and b ballots in favour of B, the probability that A is strictly ahead of B at every step of the count is $\dfrac{a-b}{a+b}$. We will assume that uniform probability is applied.
For example, if $a = 2$ and $b = 1$, then the possible tuples are AAB, ABA, BAA. Only AAB qualifies and the answer is $\dfrac{1}{3}$.

Proof. The proof is instructive because it combines uniform probability, complementary probability, and the bijection principle. Let $a + b = n$. Before we begin, we remark that the set of all n-tuples whose entries are A's and B's is finite with cardinality 2^n, and all of our sets will be subsets of this set, so we can freely speak of their cardinality. Another introductory remark is that, if $b = 0$, then the answer is 1, which matches the stated formula. So, we may assume that $b \geq 1$.
Let S be the set of n-tuples that consist of a copies of A and b copies of B. For each element of S, define the n-tuples

$$(a_1, a_2, \ldots, a_n) \text{ and } (b_1, b_2, \ldots, b_n),$$

where the first tuple is the number of votes that A has after each of the n steps of the count, and the second tuple is the number of votes that B has after each of the n steps of the count. Of importance is the tuple

$$(c_1, c_2, \ldots, c_n) = (a_1 - b_1, a_2 - b_2, \ldots, a_n - b_n),$$

since the problem is asking for the probability that $(c_1, c_2, \ldots, c_n)$ consists of only positive integers. Since exactly one ballot is counted at each step, this n-tuple satisfies the hypotheses of the discrete intermediate value. Note that we are assuming $b \geq 1$, which implies $a \geq 2$, and so $n = a + b \geq 3$.
Our first observation is that, if all of the c_i are positive, then none of the c_i are zero. Perhaps we can prove the converse, which would establish biconditionality. Actually, it is easier to tackle the contrapositive of the converse. Suppose not all of the c_i are positive. We want to show that at least one of the c_i is zero. Suppose for contradiction that none of the c_i are zero. Then, by the assumption, at least one of the c_i is negative, say c_m. And since A is the winner, c_n is positive. By the discrete intermediate value theorem, at some index k strictly between m and n, it holds that $c_k = 0$. This is a contradiction. So, now we know that it is equivalent to find the probability that none of the entries of $(c_1, c_2, \ldots, c_n)$ is zero.
Now we pull in complementary counting. We will find the probability that at least one of the entries of $(c_1, c_2, \ldots, c_n)$ is zero and subtract this from 1. Our second observation is that an element of S that reaches a tie (meaning some c_i is zero) can begin with either A or B. Let T be the set of all n-tuples of a copies of A and b copies of B that reach a tie, let T_A be the subset of T whose elements begin with A, and let T_B be the subset of T whose elements begin with B. Let S_B be the elements of S that begin with B. For every element of S_B, it is true that $c_1 = -1$ and we always have $c_n = a - b \geq 1$ for every element of S; by the discrete intermediate value theorem, a tie is then necessarily reached somewhere, so $S_B \subseteq T_B$. Of course, $T_B \subseteq S_B$ as well, so $T_B = S_B$. This is good news because S_B is easier to count than T_B, and we will do so later.
Right now, we need to deal with T_A. We will use a miraculous technique called André's reflection principle to show that T_A is in bijection with S_B: given an element of T_A, transform it by "reflecting" every A to B and every B to A from the first ballot up to and including the first ballot at which a tie is reached. This produces an element of S_B. It is easy to show that the map is surjective because, for every element of S_B, an appropriate element of T_A can be produced by reflecting every A and B up to and including the first tie (this tie must exist because $S_B = T_B$). Injectivity also holds because, if two elements of T_A disagree at some index, then their images will also disagree at the same spot. Thus, we have a bijection and the problem has been reduced to computing

$$1 - \frac{|T|}{|S|} = 1 - \frac{|T_A| + |T_B|}{|S|} = 1 - 2 \cdot \frac{|S_B|}{|S|}.$$

The cardinality of S is the multinomial coefficient $\dfrac{(a+b)!}{a! \cdot b!}$ and, similarly, the cardinality of S_B is the multinomial coefficient $\dfrac{(a+b-1)!}{a! \cdot (b-1)!}$. Dividing the latter by the former and cancelling factors yields the probability $\dfrac{b}{a+b}$, so the final answer is

$$1 - 2 \cdot \frac{|S_B|}{|S|} = 1 - 2 \cdot \frac{b}{a+b} = \frac{a-b}{a+b}.$$

□

While Theorem 9.7 might seem like a arbitrary frivolity, it can be used to find a formula for the Catalan numbers (Theorem 10.14), which is among the most significant sequences of numbers.

9.2 Almost

In probability, it is easy to refer to events as "certain" or "impossible," but there is some ambiguity in these words. The following is an exploration based on the observation that there might exist non-empty events with zero probability, and there might exist events that are not the entire sample space but have probability 1. We will introduce non-standard terminology because some of these concepts have no well-known names in the literature to the best of our knowledge.

Problem 9.8. Let n be a positive integer and $L = (A_1, A_2, \dots, A_n)$ be an n-tuple of events.

1. If $A_i \cap A_j = \varnothing$ for every pair of indices i, j such that $1 \le i < j \le n$, then we call L "pairwise disjoint," a term that came up in Kolmogorov's third axiom. More generally, L is said to be "mutually disjoint" if for any integer k such that $2 \le k \le n$ and any k indices $1 \le i_1 < i_2 < \cdots < i_k \le n$, it is true that
$$A_{i_1} \cap A_{i_2} \cap \cdots \cap A_{i_k} = \varnothing.$$
Show that L is pairwise disjoint if and only if L is mutually disjoint, even though the latter seems to be a stronger criterion than the former. As such, we will cease to use these two terms, and instead say that L is **mutually exclusive** if it has these equivalent attributes.

2. An event A is said to occur **almost never** if $\mathbb{P}(A) = 0$. Show that $A_i \cap A_j$ occurs almost never for every pair of indices i, j such that $1 \le i < j \le n$ if and only if it is true, for every integer k such that $2 \le k \le n$ and every k indices $1 \le i_1 < i_2 < \cdots < i_k \le n$, that $A_{i_1} \cap A_{i_2} \cap \cdots \cap A_{i_k}$ occurs almost never. If L has these equivalent attributes, then we say that L is **almost mutually exclusive**.

3. Show that if L is mutually exclusive, then L is almost mutually exclusive.

4. An event A is said to occur **almost surely** if $\mathbb{P}(A) = 1$. We say that L is **collectively exhaustive** if $A_1 \cup A_2 \cup \cdots \cup A_n = \Omega$. If $A_1 \cup A_2 \cup \cdots \cup A_n$ occurs almost surely, then we say that L is **almost collectively exhaustive**. Show that if L is collectively exhaustive, then it is almost collectively exhaustive.

5. A finite probability space is said to be **positive** if, for all events A, if $\mathbb{P}(A) = 0$ then $A = \varnothing$. Suppose we are working in a positive finite probability space. Show that:
   - If L is almost mutually exclusive then L is mutually exclusive.
   - If L is almost collectively exhaustive then L is collectively exhaustive.

   It not necessary to prove, but it is easy to see that these two statements cannot be asserted if our finite probability space is not positive, at least not without more information about L.

Definition 9.9. We will use the terminology of Problem 9.8. Let n be a positive integer and L be an n-tuple of events. If L is both mutually exclusive and collectively exhaustive, then L is said to be a **partition** of Ω. If L is both almost mutually exclusive and almost collectively exhaustive, then L is said to be an **almost-partition** of Ω.

A partition of a sample space has the same underlying concept as a generalized partition of a set (Definition 1.30) and a partition of a non-empty set (Definition 7.22), but there are differences in our chosen definitions. For example, one is a tuple, one is a multiset, and one is a set.

Problem 9.10. As an exercise of little importance, prove that $\mathbb{P}(A \cup B) = \mathbb{P}(A) \cdot \mathbb{P}(B)$ if and only if A and B both occur almost never or both occur almost surely. The equation that replaces the union with intersection is of greater significance (see Definition 9.15).

Theorem 9.11 (Boole's inequality)**.** Suppose n is a positive integer and that we have an n-tuple of events $L = (A_1, A_2, \dots, A_n)$. Show that
$$\mathbb{P}\left(\bigcup_{i=1}^{n} A_i\right) \le \sum_{i=1}^{n} \mathbb{P}(A_i),$$
with equality holding if and only if L is almost mutually exclusive.

Proof. We will perform a proof by induction on $n \geq 1$. In the base case $n = 1$, it is clearly true that $\mathbb{P}(A_1) \leq \mathbb{P}(A_1)$. Equality always holds, and a 1-tuple of events (A_1) is vacuously almost mutually exclusive. Suppose the result is true for some integer $n \geq 1$. Let

$$(A_1, A_2, \ldots, A_n, A_{n+1})$$

be an $(n+1)$-tuple of events. Let $B = A_1 \cup A_2 \cup \cdots \cup A_n$. By the probabilistic principle of inclusion-exclusion for two events and the first Kolmogorov axiom,

$$\begin{aligned} \mathbb{P}\left(\bigcup_{i=1}^{n+1} A_i\right) &= \mathbb{P}(B \cup A_{n+1}) = \mathbb{P}(B) + \mathbb{P}(A_{n+1}) - \mathbb{P}(B \cap A_{n+1}) \\ &\leq \mathbb{P}(B) + \mathbb{P}(A_{n+1}) \leq \sum_{i=1}^{n} \mathbb{P}(A_i) + \mathbb{P}(A_{n+1}), \end{aligned}$$

where we used the induction hypothesis in the last step. All we have to do now is establish the equality condition. Equality holds if and only if each inequality stated is actually an equality. So equality holds if and only if

$$\mathbb{P}(B \cap A_{n+1}) = 0 \text{ and } \mathbb{P}(B) = \sum_{i=1}^{n} \mathbb{P}(A_i).$$

Assuming the first criterion,

$$\begin{aligned} 0 &= \mathbb{P}(B \cap A_{n+1}) \\ &= \mathbb{P}((A_1 \cup A_2 \cup \cdots \cup A_n) \cap A_{n+1}) \\ &= \mathbb{P}((A_1 \cap A_{n+1}) \cup (A_2 \cap A_{n+1}) \cup \cdots \cup (A_n \cap A_{n+1})). \end{aligned}$$

For each index $1 \leq i \leq n$, monotonicity and the first Kolmogorov axiom and the preceding fact yield

$$0 \leq \mathbb{P}(A_i \cap A_{n+1}) \leq \mathbb{P}((A_1 \cap A_{n+1}) \cup (A_2 \cap A_{n+1}) \cup \cdots \cup (A_n \cap A_{n+1})) = 0.$$

Thus, the first criterion implies that $\mathbb{P}(A_i \cap A_{n+1}) = 0$ for each index i. Conversely, if $\mathbb{P}(A_i \cap A_{n+1}) = 0$ for every i, then by applying the induction hypothesis to the n-tuple

$$(A_1 \cap A_{n+1}, A_2 \cap A_{n+1}, \ldots, A_n \cap A_{n+1}),$$

we get

$$0 = \sum_{i=1}^{n} \mathbb{P}(A_i \cap A_{n+1}) \geq \mathbb{P}\left(\bigcup_{i=1}^{n} (A_i \cap A_{n+1})\right) = \mathbb{P}(B \cap A_{n+1}) \geq 0.$$

By antisymmetry, this forces $\mathbb{P}(B \cap A_{n+1}) = 0$. Therefore, the first criterion in the equality condition is equivalent to it being true that $\mathbb{P}(A_i \cap A_{n+1}) = 0$ for every index $1 \leq i \leq n$.
As for the second criterion, applying the induction hypothesis to the n-tuple $(A_1, A_2, \ldots, A_n)$ says that

$$\mathbb{P}(B) = \sum_{i=1}^{n} \mathbb{P}(A_i)$$

if and only if $(A_1, A_2, \ldots, A_n)$ is almost mutually exclusive. Put together, the two criteria yield that equality holds if and only if

$$(A_1, A_2, \ldots, A_n, A_{n+1})$$

is almost mutually exclusive. □

Elementary sources sometimes treat a different kind of probability space, in a context called continuous or geometric probability. There, the sample space is a subset of the Euclidean plane or Euclidean space, and the probability distribution function is the ratio of the area or volume of an event to the area or volume of the sample space. We have not written about such probability spaces because writing a proper treatment about them would lead us down to the advanced topic of of measure theory. For the curious reader, the starting point is the issue of strange subsets of the plane that cannot be assigned an area. We also recommend studying Bertrand's paradox in probability, which provides an interesting glimpse into the question of a probability space being well-defined.

9.3 Conditional Probability

As before, we let the event space be Ω, the atomic probability distribution be p, and the probability distribution be $\mathbb{P}$, unless otherwise specified.

Definition 9.12. If B is an event with non-zero probability, then for any event A, the **conditional probability** of A given B is denoted by and defined as

$$\mathbb{P}(A \mid B) = \frac{\mathbb{P}(A \cap B)}{\mathbb{P}(B)}.$$

Intuitively, the probability of an event A, conditional upon some specific event B occurring, is the new probability of A upon contracting the sample space from Ω to B.

Example. The case that is least amenable to easy computation is when A and B have a non-empty intersection but neither is contained inside the other and they are not disjoint. The other cases can be computed as

$$\mathbb{P}(A \mid B) = \frac{\mathbb{P}(A \cap B)}{\mathbb{P}(B)} = \begin{cases} \frac{\mathbb{P}(A)}{\mathbb{P}(B)} & \text{if } A \subseteq B \\ 1 & \text{if } B \subseteq A \\ 0 & \text{if } A \cap B = \varnothing \end{cases}.$$

These can be easily visualized using Venn diagrams with two circles.

Problem 9.13. In general, it is not true that conditional probability is symmetrical, meaning we cannot always claim that the probability of A given B is the same as the probability of B given A. Assuming $\mathbb{P}(A) \neq 0$ and $\mathbb{P}(B) \neq 0$, show that this symmetry holds if and only if $\mathbb{P}(A) = \mathbb{P}(B)$.

Theorem 9.14. Let B be a event with non-zero probability. Then the function

$$\begin{aligned} \mathbb{P}(\cdot \mid B) : \mathcal{P}(\Omega) &\to \mathbb{R}_{\geq 0} \\ A &\mapsto \mathbb{P}(A \mid B) \end{aligned}$$

is a probability distribution in the sense that there exists an atomic probability distribution $p_B : \Omega \to \mathbb{R}_{\geq 0}$ that gives rise to $\mathbb{P}(\cdot \mid B)$. As such, the Kolmogorov axioms (Theorem 9.4) and their implications (Theorem 9.5) hold for $\mathbb{P}(\cdot \mid B)$.

Proof. As we mentioned at the beginning, the motivating idea is to contract the sample space from Ω to B, so we define $p_B : \Omega \to \mathbb{R}_{\geq 0}$ by

$$p_B(\omega) = \begin{cases} \frac{\mathbb{P}(\{\omega\})}{\mathbb{P}(B)} & \text{if } \omega \in B, \\ 0 & \text{if } \omega \in \overline{B} \end{cases}.$$

Then we can prove that p_B satisfies the requirement of being an atomic probability distribution because its values on Ω add up to 1 as follows. Using the fact that B and $\overline{B}$ form a partition of Ω,

$$\sum_{\omega \in \Omega} p_B(\omega) = \sum_{\omega \in B} p_B(\omega) + \sum_{\omega \in \overline{B}} p_B(\omega) = \sum_{\omega \in B} \frac{\mathbb{P}(\{\omega\})}{\mathbb{P}(B)} + \sum_{\omega \in \overline{B}} 0 = \frac{1}{\mathbb{P}(B)} \cdot \sum_{\omega \in B} p(\omega) = \frac{1}{\mathbb{P}(B)} \cdot \mathbb{P}(B) = 1.$$

Finally, we must show that p_B gives rise to $\mathbb{P}(\cdot \mid B)$ in the following sense. It is already clear that $\mathbb{P}(\cdot \mid B)$ is a function from $\mathcal{P}(\Omega)$ to $\mathbb{R}_{\geq 0}$. We have to show that for each $A \in \mathcal{P}(\Omega)$,

$$\mathbb{P}(A \mid B) = \sum_{\omega \in A} p_B(\omega).$$

Indeed, by splitting A according to which of its elements lie in B or $\overline{B}$, we get

$$\sum_{\omega \in A} p_B(\omega) = \sum_{\substack{\omega \in A \\ \omega \in B}} p_B(\omega) + \sum_{\substack{\omega \in A \\ \omega \in \overline{B}}} p_B(\omega) = \sum_{\substack{\omega \in A \\ \omega \in B}} \frac{\mathbb{P}(\{\omega\})}{\mathbb{P}(B)} + \sum_{\substack{\omega \in A \\ \omega \in \overline{B}}} 0 = \frac{1}{\mathbb{P}(B)} \cdot \sum_{\omega A \cap B} p(\omega) = \frac{\mathbb{P}(A \cap B)}{\mathbb{P}(B)} = \mathbb{P}(A \mid B).$$

□

Conditional probability leads to the following question: Given that an event B occurs, what are the events A whose probabilities remain unperturbed by this knowledge?

Definition 9.15. There are several increasingly general notions of independence.

1. Two events A and B are **independent** if
$$\mathbb{P}(A \cap B) = \mathbb{P}(A) \cdot \mathbb{P}(B).$$
This is inspired by the fact that, if B has non-zero probability, then the condition is equivalent to
$$\mathbb{P}(A \mid B) = \mathbb{P}(A),$$
which says that the knowledge of B occurring does not affect the probability of A. Interestingly, if the probability of A is non-zero, then the definition is also equivalent to $\mathbb{P}(B \mid A) = \mathbb{P}(B)$, showcasing the symmetry of the definition. The first definition given is usually preferable because it immediately shows that A and B are symmetric in the relation of independence and because it does not rely on any probability being non-zero.

2. If n is a positive integer and $L = (A_1, A_2, \ldots, A_n)$ is an n-tuple of events, then L is said to be **pairwise independent** if, for all pairs of indices i, j such that $1 \leq i < j \leq n$, the events A_i and A_j are independent.

3. In terms of n and L again, L is said to be **mutually independent** if, for every non-empty subset $I \subseteq [n]$,
$$\mathbb{P}\left(\bigcap_{i \in I} A_i\right) = \prod_{i \in I} \mathbb{P}(A_i).$$
Note that this must hold for *all* non-empty subsets I of $[n] = \{1, 2, \ldots, n\}$ in order to satisfy the definition of mutual independence.

It is clear that mutual independence implies pairwise independence, but there exist counterexamples for the converse. As such, unlike how the equivalence between being pairwise disjoint and being mutually disjoint allowed us to relabel them both as being mutually exclusive (see Problem 9.8), we will need to retain separate labels for these two kinds of independence.

Problem 9.16. Solve the following problems about independence:

1. Prove that an event A is independent of itself if and only if A occurs almost never or A occurs almost surely.

2. Find an example of a triple of events (A, B, C) of events such that
$$\mathbb{P}(A \cap B \cap C) = \mathbb{P}(A) \cdot \mathbb{P}(B) \cdot \mathbb{P}(C),$$
without it being true that (A, B, C) is mutually independent.

Problem 9.17. Prove that events A and B are simultaneously independent and almost mutually exclusive if and only if at least one of A or B occurs almost never.

To get a good sense of what the following theorem says, the reader might find it instructive to write out the equation for $n = 4$ without product notation.

Theorem 9.18 (Chain rule for events)**.** By the definition of conditional probability, if A and B are events such that $\mathbb{P}(A) \neq 0$, then
$$\mathbb{P}(A \cap B) = \mathbb{P}(A) \cdot \mathbb{P}(B \mid A).$$
The following is a generalization of this idea of expressing the probability of an intersection of events as the product of conditional probabilities. If $n \geq 2$ is an integer and $(A_1, A_2, \ldots, A_n)$ is an n-tuple of events such that $\mathbb{P}\left(\bigcap_{i=1}^{k-1} A_i\right) \neq 0$ for every integer k satisfying $2 \leq k \leq n$, then
$$\mathbb{P}\left(\bigcap_{i=1}^{n} A_i\right) = \mathbb{P}(A_1) \cdot \prod_{k=2}^{n} \mathbb{P}\left(A_k \,\middle|\, \bigcap_{i=1}^{k-1} A_i\right).$$

Proof. We can prove this by induction on $n \geq 2$. As stated at the beginning of the theorem, the $n = 2$ case holds by the definition of conditional probability. With the base case established, suppose the result holds for some integer $n \geq 2$. Let $(A_1, A_2, \dots, A_n, A_{n+1})$ be an $(n+1)$-tuple of events such that $\mathbb{P}\left(\bigcap_{i=1}^{k-1} A_i\right) \neq 0$ for $k = 2, 3, \dots, n, n+1$. By the definition of conditional probability,

$$\mathbb{P}\left(\bigcap_{i=1}^{n+1} A_i\right) = \mathbb{P}\left(\left(\bigcap_{i=1}^{n} A_i\right) \cap A_{n+1}\right) = \mathbb{P}\left(\bigcap_{i=1}^{n} A_i\right) \cdot \mathbb{P}\left(A_{n+1} \,\middle|\, \bigcap_{i=1}^{n} A_i\right).$$

The inductive hypothesis is applicable to the n-tuple of events $(A_1, A_2, \dots, A_n)$, so we can replace $\mathbb{P}\left(\bigcap_{i=1}^{n} A_i\right)$ in the equation with $\mathbb{P}(A_1) \cdot \prod_{k=2}^{n} \mathbb{P}\left(A_k \,\middle|\, \bigcap_{i=1}^{k-1} A_i\right)$, which completes the induction. Less formally, we can simply expand out the product and use telescoping between the numerators and denominators. □

Problem 9.19. Let $n \geq 2$ be an integer and

$$L = (A_1, A_2, \dots, A_n)$$

be an n-tuple of events.

1. Show that, if L is mutually independent, then, for any index $1 \leq i \leq n$ and non-empty subset
$$J \subseteq [n] \backslash \{i\} = \{1, 2, \dots, n\} \backslash \{i\},$$
if $\mathbb{P}\left(\bigcap_{j \in J} A_j\right) \neq 0$, then $\mathbb{P}\left(A_i \,\middle|\, \bigcap_{j \in J} A_j\right) = \mathbb{P}(A_i)$.

2. As somewhat of a converse to the first part, show that, if it is true, for every integer k such that $2 \leq k \leq n$ and every set of k indices $\{i_1, i_2, \dots, i_k\}$ such that $1 \leq i_1 < i_2 < \cdots < i_k \leq n$, that $\mathbb{P}\left(\bigcap_{j=1}^{k-1} A_{i_j}\right) \neq 0$ and $\mathbb{P}\left(A_{i_k} \,\middle|\, \bigcap_{j=1}^{k-1} A_{i_j}\right) = \mathbb{P}(A_{i_k})$, then L is mutually independent. In words, if every set of at least two events in L satisfies the assertion that the event with the highest index is independent of the intersection of the rest of the events in the set, then L is mutually independent. Note that an alternate stronger hypothesis that is easier to state is that every set of at least two events in L satisfies the assertion that *any* of the events in the set is independent of the intersection of the other events in the set, but we can get away with the weaker stated hypothesis.

This exercise provides a meaning for mutual independence in terms of conditional probability, though the original multiplicative definition in Definition 9.15 is preferable because the it does not rely on any probabilities being non-zero.

In probability, the union of events corresponds to the word "or" because at least one of the events occurs; as a side note, this is different from the "exclusive or," which requires exactly one of the events to occur. In contrast, the intersection of events corresponds to the word "and" because all of the events must occur. Our computational tools for the union of events are the third Kolmogorov axiom (Theorem 9.4) and its generalization, Boole's inequality (Theorem 9.11). For the intersection of events, we now have the chain rule for events (Theorem 9.18). Intersections and unions come together in the next result.

Theorem 9.20 (Law of total probability). Suppose A is an event, and n is a positive integer, and we have an almost-partition $(B_1, B_2, \dots, B_n)$ of Ω (Definition 9.9) such that none of the B_i have zero probability. Then

$$\mathbb{P}(A) = \sum_{i=1}^{n} \mathbb{P}(A \cap B_i) = \sum_{i=1}^{n} \mathbb{P}(A \mid B_i) \cdot \mathbb{P}(B_i).$$

A particular instance of this theorem corresponds to $n = 2$ and the events $(B, \overline{B})$, where B is taken to occur neither almost never nor almost surely. Then

$$\mathbb{P}(A) = \mathbb{P}(A \cap B) + \mathbb{P}(A \cap \overline{B}) = \mathbb{P}(A \mid B) \cdot \mathbb{P}(B) + \mathbb{P}(A \mid \overline{B}) \cdot \mathbb{P}(\overline{B}).$$

Proof. For each index i, the definition of conditional probability immediately states that

$$\mathbb{P}(A \cap B_i) = \mathbb{P}(A \mid B_i) \cdot \mathbb{P}(B_i).$$

So all we need to do is prove the first equality

$$\mathbb{P}(A) = \sum_{i=1}^{n} \mathbb{P}(A \cap B_i).$$

By applying Boole's inequality to the n-tuple of events

$$(A \cap B_1, \ldots, A \cap B_n),$$

we get

$$\sum_{i=1}^{n} \mathbb{P}(A \cap B_i) \geq \mathbb{P}\left(\bigcup_{i=1}^{n} (A \cap B_i)\right) = \mathbb{P}\left(A \cap \left(\bigcup_{i=1}^{n} B_i\right)\right).$$

By the two-event probabilistic principle of inclusion-exclusion,

$$\mathbb{P}\left(A \cap \left(\bigcup_{i=1}^{n} B_i\right)\right) = \mathbb{P}(A) + \mathbb{P}\left(\bigcup_{i=1}^{n} B_i\right) - \mathbb{P}\left(A \cup \left(\bigcup_{i=1}^{n} B_i\right)\right).$$

Since $(B_1, B_2, \ldots, B_n)$ is almost collectively exhaustive, we can use monotonicity to get

$$1 = \mathbb{P}\left(\bigcup_{i=1}^{n} B_i\right) \leq \mathbb{P}\left(A \cup \left(\bigcup_{i=1}^{n} B_i\right)\right) \leq 1,$$

so $\mathbb{P}\left(\bigcup_{i=1}^{n} B_i\right) = \mathbb{P}\left(A \cup \left(\bigcap_{i=1}^{n} B_i\right)\right) = 1$. Thus, we get the inequality

$$\sum_{i=1}^{n} \mathbb{P}(A \cap B_i) \geq \mathbb{P}(A),$$

with the equality condition for Boole's inequality stating that equality holds if and only if $(A \cap B_1, \ldots, A \cap B_n)$ is almost mutually exclusive. We will show that this equality condition is implied by $(B_1, B_2, \ldots, B_n)$ being almost mutually exclusive. Indeed, for any pair of indices i, j such that $1 \leq i < j \leq n$, since $\mathbb{P}(B_i \cap B_j) = 0$, it holds that

$$0 \leq \mathbb{P}((A \cap B_i) \cap (A \cap B_j)) = \mathbb{P}(A \cap (B_i \cap B_j)) \leq \mathbb{P}(B_i \cap B_j) = 0.$$

Thus, $A \cap B_i$ and $A \cap B_j$ are almost mutually exclusive, as desired. □

Problem 9.21. Suppose A is an event with non-zero probability and n is a positive integer, and we have an almost-partition $(B_1, B_2, \ldots, B_n)$ of Ω. Show that

$$\sum_{i=1}^{n} \mathbb{P}(B_i \mid A) = 1.$$

In particular, for any event A with non-zero probability and any event B,

$$\mathbb{P}(B \mid A) + \mathbb{P}(\overline{B} \mid A) = 1.$$

Our final theorem of the section is an important result that one mathematician described as being "to the theory of probability what the Pythagorean theorem is to geometry."

Theorem 9.22 (Bayes's rule)**.** For events A and B that both have non-zero probability, it holds that

$$\mathbb{P}(B \mid A) = \frac{\mathbb{P}(A \mid B) \cdot \mathbb{P}(B)}{\mathbb{P}(A)}.$$

There is an extended variation as follows. If A is an event with non-zero probability, n is a positive integer and $(B_1, B_2, \ldots, B_n)$ is an almost-partition of Ω consisting of events with non-zero probability, then for each index k satisfying $1 \leq k \leq n$, it holds that

$$\mathbb{P}(B_k \mid A) = \frac{\mathbb{P}(A \mid B_k) \cdot \mathbb{P}(B_k)}{\sum_{i=1}^{n} \mathbb{P}(A \mid B_i) \cdot \mathbb{P}(B_i)}.$$

In particular, if $n = 2$ and the list of events is $(B, \overline{B})$ for some event B that occurs neither almost never nor almost surely, then

$$\mathbb{P}(B \mid A) = \frac{\mathbb{P}(A \mid B) \cdot \mathbb{P}(B)}{\mathbb{P}(A \mid B) \cdot \mathbb{P}(B) + \mathbb{P}(A \mid \overline{B}) \cdot \mathbb{P}(\overline{B})}.$$

Proof. The first formula follows from the definition of conditional probability because

$$\frac{\mathbb{P}(A \mid B) \cdot \mathbb{P}(B)}{\mathbb{P}(A)} = \frac{\mathbb{P}(A \cap B) \cdot \mathbb{P}(B)}{\mathbb{P}(B) \cdot \mathbb{P}(A)} = \frac{\mathbb{P}(A \cap B)}{\mathbb{P}(A)} = \mathbb{P}(B \mid A).$$

For the extended variation, applying the first formula to $B = B_k$ yields

$$\mathbb{P}(B_k \mid A) = \frac{\mathbb{P}(A \mid B_k) \cdot \mathbb{P}(B_k)}{\mathbb{P}(A)}.$$

Then we replace the denominator using the law of total probability (Theorem 9.20), which completes the proof. The final binary formula is an immediate consequence of the extended variation. □

Our writing on conditional probability would be incomplete without mention of the Monty hall problem. It is a famous question that reaches such a counterintuitive answer that the conclusion remained controversial for some time. However, the mathematics has been backed up by repeated experiments. [5]

Example 9.23 (Monty hall problem)**.** You are on a game show. There are three closed doors labelled 1, 2, 3. Behind two of the doors are goats and behind the remaining door is a brand new car. You get to choose a door and your reward is what lies behind the door that you choose. You pick a door. The game show's host then opens a different door to reveal a goat. You are then given the chance to switch to the only other door that remains closed or to stick to your originally chosen door. Assuming your goal is to receive the car, what do you do?

Solution. We have seen attempted solutions to this problem that use Bayes's rule, but the reasoning usually seemed fuzzy to us because, as with many problems and solutions in probability, the sample space was not made explicit. In this proof, we have done our utmost to create an airtight argument, in particular by clearly defining the sample space.

Formally, the sample space consists of triples (A, B, C) where each of A, B, C is an element of $\{1, 2, 3\}$ (potentially with repeats) satisfying the following criteria. In any particular triple (A, B, C), A is the door behind which the car lies, B is the door you initially chose, and C is the door that the host opened (the C matters because it turns out to provide extra information to the player in the decision to change the original choice). There are $3^3 = 27$ such triples, but not all are elements of Ω because the host must choose C such that $C \neq A$ and $C \neq B$. We have drawn a tree diagram of the possible triples below in order to show that there are 12 such triples in the sample space (we will prove this momentarily). We want to find the probabilities of the following two sets:

$$\Omega_1 = \{(A, B, C) \in \Omega : A = B\},$$
$$\Omega_2 = \{(A, B, C) \in \Omega : A \neq B\}.$$

The probability of the first set is the probability that staying with your initially chosen door is the correct choice, and the probability of the second set is the probability that switching doors is the correct choice. The two sets are complements, so finding the probability of one should yield the other but we will compute them independently so that no doubt is left.
Here is the critical observation: There are $3 \cdot 1 = 3$ ways in which $A = B$ and for each such possibility, there are 2 choices that the host has for C, yielding $|\Omega_1| = 3 \cdot 2 = 6$. Similarly, there are $3 \cdot 2 = 6$ ways in which $A \neq B$ and for each such possibility, there is just 1 choice that the host has for C, yielding $|\Omega_2| = 6 \cdot 1 = 6$. At this point, it might seem like the $\mathbb{P}(\Omega_1) = \mathbb{P}(\Omega_2) = \frac{1}{2}$, but possibility does not translate to probability. Less cryptically, we cannot assume that this probability space has the uniform probability.

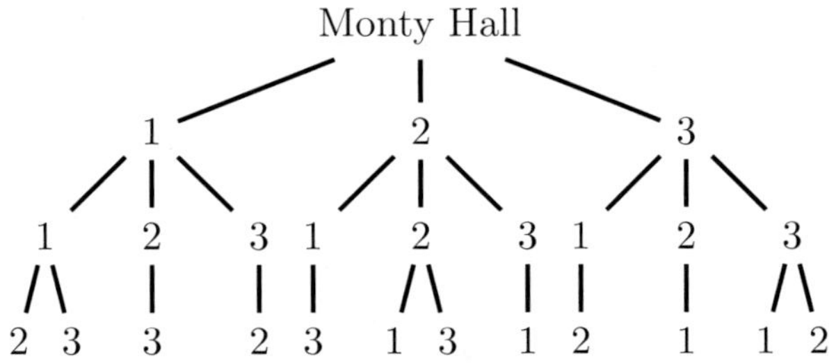

Similar to how we traversed the tree of possibilities in the dependent multiplication principle (Theorem 3.9), we will "travel through" the coordinates of (A, B, C) from A to B to C using the chain rule for events (Theorem 9.18). To this end, we define the following sets:

- $S(a)$ is the set of events (A, B, C) with $A = a$
- $S(a, b)$ is the set of events (A, B, C) with $A = a$ and $B = b$
- $S(a, b, c)$ is the set of events (A, B, C) with $A = a$ and $B = b$ and $C = c$

By the chain rule for events,

$$\begin{aligned}\mathbb{P}(\{(a,b,c)\}) &= \mathbb{P}(S(a) \cap S(a,b) \cap S(a,b,c)) \\ &= \mathbb{P}(S(a)) \cdot \mathbb{P}(S(a,b) \mid S(a)) \cdot \mathbb{P}(S(a,b,c) \mid S(a,b)) \\ &= \frac{1}{3} \cdot \frac{1}{3} \cdot \mathbb{P}(S(a,b,c) \mid S(a,b)).\end{aligned}$$

Computation of $\mathbb{P}(S(a,b,c) \mid S(a,b))$ depends on whether $a = b$ or $a \neq b$. If $a = b$, then the host has two choices and so $\mathbb{P}(S(a,b,c) \mid S(a,b)) = \frac{1}{2}$. If $a \neq b$, then the host has one choice and so $\mathbb{P}(S(a,b,c) \mid S(a,b)) = 1$. Since $|\Omega_1| = |\Omega_2| = 6$ and we have shown that the probability is uniform within each of Ω_1 and Ω_2 individually,

$$\begin{aligned}\mathbb{P}(\Omega_1) &= 6 \cdot \frac{1}{3} \cdot \frac{1}{3} \cdot \frac{1}{2} = \frac{1}{3}, \\ \mathbb{P}(\Omega_2) &= 6 \cdot \frac{1}{3} \cdot \frac{1}{3} \cdot 1 = \frac{2}{3}.\end{aligned}$$

Therefore, it is in your interest to switch doors. This will not always result in winning, but it will switch the probability of winning from $\frac{1}{3}$ to $\frac{2}{3}$, which doubles your chances. □

9.4 Expected Value

Definition 9.24. A **random variable** on a finite probability space is any real-valued function with the sample space as its domain. So it is any function $f : \Omega \to \mathbb{R}$.

Example. If A is an event, then the **indicator variable** of A is the function $I_A : \Omega \to \mathbb{R}$, which is defined as

$$I_A(x) = \begin{cases} 1 & \text{if } x \in A \\ 0 & \text{if } x \in \overline{A} \end{cases}.$$

Definition 9.25. The **expected value** of a random variable $f : \Omega \to \mathbb{R}$ is denoted by and defined as

$$\mathbb{E}(f) = \sum_{\omega \in \Omega} p(\omega) f(\omega).$$

In words, this is the sum of the weighted outputs of the random variable on each element of the sample space, where each output is weighed according to the probability of the element of the sample space.

Example. The expected value of the indicator variable I_A is

$$\begin{aligned} \mathbb{E}(I_A) &= \sum_{\omega \in \Omega} p(\omega) I_A(\omega) = \sum_{\omega \in A} p(\omega) I_A(\omega) + \sum_{\omega \in \overline{A}} p(\omega) I_A(\omega) \\ &= \sum_{\omega \in A} p(\omega) \cdot 1 + \sum_{\omega \in \overline{A}} p(\omega) \cdot 0 = \sum_{\omega \in A} p(\omega) = \mathbb{P}(A). \end{aligned}$$

Another example is that if f is any random variable on a probability space that has the uniform probability, then

$$\mathbb{E}(f) = \frac{1}{|\Omega|} \cdot \sum_{\omega \in \Omega} f(\omega),$$

which is akin to the average of the values of f on Ω.

Example 9.26. Let the sample space be

$$\Omega = \{HH, HT, TH, TT\},$$

which is the set of all outcomes of two consecutive coin tosses using a fair coin; of course, H denotes heads and T denotes tails. Determine the expected number of heads.

Solution. Using the uniform probability (since the coin is fair) and letting the random variable be h,

$$\mathbb{E}(h) = \frac{1}{4}(2 + 1 + 1 + 0) = 1.$$

Thus, the expected number of heads is 1. □

Problem 9.27. For non-negative integers n, determine the expected number of elements of a random subset of $[n] = \{1, 2, \ldots, n\}$.

Example 9.28 (Binomial distribution)**.** An unfair coin has a probability p of tossing heads and a probability $q = 1 - p$ of tossing tails. For a positive integer n and an integer k such that $0 \leq k \leq n$, determine the probability that, if the coin is tossed n times, then exactly k tosses are heads. Use this to find the expected number of tosses that are heads if the coin is tossed n times.

Solution. To get exactly k heads out of n tosses, there are $\binom{n}{k}$ ways in ways in which it can happen, since the tosses occur at distinct times (like indices). In each such successful case, each of the k heads have probability p and each of the tails have probability $q = 1 - p$. Since the tosses are all independent of each other,

$$\mathbb{P}(X = k) = \binom{n}{k} p^k q^{n-k},$$

where X is the random variable that represents the number of heads out of n tosses.

As a result, the expected number of heads is

$$\begin{aligned}\mathbb{E}[X] &= \sum_{k=0}^{n} k \cdot \mathbb{P}(X = k) = \sum_{k=0}^{n} k \binom{n}{k} p^k q^{n-k} \\ &= \sum_{k=0}^{n} k \cdot \frac{n}{k} \cdot \binom{n-1}{k-1} p^k q^{n-k} = np \sum_{k=0}^{n} \binom{n-1}{k-1} p^{k-1} q^{(n-1)-(k-1)} \\ &= np(p+q)^{n-1} = np \cdot 1^{n-1} = np,\end{aligned}$$

where we used the binomial theorem to evaluate the sum. □

Example 9.29 (Hypergeometric distribution). Let N be a positive integer and K be an integer such that $0 \le K \le N$. There are N distinct birds in a barn, out of which K are distinct owls and $N - K$ are distinct crows. For positive integers k, n such that $0 \le k \le n \le N$, if n birds are randomly selected, then find the probability that exactly k of them are owls. Use this to find the expected number of birds that are owls out of the n randomly chosen birds.

Solution. If X is the random variable that assigns to a selection of n birds the number of them that are owls, then there are $\binom{K}{k}$ ways to select k owls, $\binom{N-K}{n-k}$ ways to select the rest of them as crows, and there are $\binom{N}{n}$ overall possible random selections of n birds. As a result,

$$\mathbb{P}(X = k) = \frac{\binom{K}{k}\binom{N-K}{n-k}}{\binom{N}{n}}.$$

As a consequence, we find that

$$\begin{aligned}\mathbb{E}[X] &= \sum_{k=0}^{n} k \cdot \mathbb{P}(X = k) = \sum_{k=0}^{n} k \cdot \frac{\binom{K}{k}\binom{N-K}{n-k}}{\binom{N}{n}} \\ &= \frac{1}{\binom{N}{n}} \sum_{k=0}^{n} k \cdot \frac{K}{k} \cdot \binom{K-1}{k-1}\binom{N-K}{n-k} = \frac{K}{\binom{N}{n}} \sum_{k=0}^{n} \binom{K-1}{k-1}\binom{(N-1)-(K-1)}{(n-1)-(k-1)} \\ &= \frac{K}{\frac{N}{n}\binom{N-1}{n-1}} \cdot \binom{N-1}{n-1} = \frac{nK}{N},\end{aligned}$$

where we used the Vandermonde identity (Theorem 5.9) to evaluate the sum. □

Clearly, we need a less *ad hoc* way of evaluating expected values, which is where its linearity comes in.

Theorem 9.30 (Linearity of expected value). Expectation is linear over random variables, meaning for any real constant c and any random variables f and g with the same domain Ω,

$$\mathbb{E}(cf + g) = c \cdot \mathbb{E}(f) + \mathbb{E}(g).$$

As a consequence,

$$\mathbb{E}(c_1 f_1 + c_2 f_2 + \cdots + c_n f_n) = c_1 \mathbb{E}(f_1) + c_2 \mathbb{E}(f_2) + \cdots + c_n \mathbb{E}(f_n),$$

for constants c_i, random variables f_i with the common domain Ω, and some positive integer n

Proof. By the definition of expected value,

$$\begin{aligned}\mathbb{E}(cf + g) &= \sum_{\omega \in \Omega} p(\omega)(cf + g)(\omega) = \sum_{\omega \in \Omega} p(\omega)(cf(\omega) + g(\omega)) \\ &= c \cdot \sum_{\omega \in \Omega} p(\omega) f(\omega) + \sum_{\omega \in \Omega} p(\omega) g(\omega) = c \cdot \mathbb{E}(f) + \mathbb{E}(g).\end{aligned}$$

The more general result for a linear combination of functions follows from a straightforward induction argument. □

There is a related concept of summing random variables which do not necessarily have the same domains and this definition also admits a linearity result, but we will not write about it because it involves the construction of a new probability space.

Problem 9.31. Suppose n is a positive integer. Recall that, given a bijection $\sigma : [n] \to [n]$, an element of the domain $i \in [n]$ is said to be a fixed point of σ if $\sigma(i) = i$. If the probability space consists of the sample space of all bijections $\sigma : [n] \to [n]$ with the uniform probability, then determine the expected number of fixed points. To be clear, this is the expected value of the random variable that sends each bijection to the number of fixed points of the bijection. As a reminder, we found in Problem 3.14 that the set of bijections $\psi : [m] \to [m]$ is finite with cardinality $m!$ for each positive integer m.

Definition 9.32. For a random variable f and a real number α, the notation $(f = \alpha)$ denotes the event defined by the preimage

$$f^{-1}(\alpha) = \{\omega \in \Omega : f(\omega) = \alpha\}.$$

Similarly, the notation $(f \leq \alpha)$ denotes the event defined by the preimage

$$f^{-1}((\infty, \alpha]) = \{\omega \in \Omega : f(\omega) \leq \alpha\}.$$

We leave it to the reader to write out the natural definitions of $(f < \alpha)$, $(f \geq \alpha)$ and $(f > \alpha)$.

Theorem 9.33 (Markov's inequality)**.** If f is a random variable whose output is always non-negative and α is a positive real number, then

$$\mathbb{P}(f \geq \alpha) \leq \frac{\mathbb{E}(f)}{\alpha}.$$

Proof. The main idea is to distribute the atomic events $\omega \in \Omega$ according to whether $\omega \in (f \geq \alpha)$ or $\omega \in (f < \alpha)$, and to simply drop the latter set. Using this technique, we get

$$\begin{aligned}\mathbb{E}(f) &= \sum_{\omega \in \Omega} p(\omega) f(\omega) = \sum_{\omega \in (f \geq \alpha)} p(\omega) f(\omega) + \sum_{\omega \in (f < \alpha)} p(\omega) f(\omega) \\ &\geq \sum_{\omega \in (f \geq \alpha)} p(\omega) f(\omega) \geq \sum_{\omega \in (f \geq \alpha)} \alpha f(\omega) = \alpha \cdot \mathbb{P}(f \geq \alpha).\end{aligned}$$

The desired result follows from dividing by α. □

While expected value is linear in general (Theorem 9.30), the multiplicative property $\mathbb{E}(fg) = \mathbb{E}(f) \cdot \mathbb{E}(g)$ is not always true. Thankfully, multiplicativity does hold under the assumption that the two random variables are independent, which is a notion that we will define now. Although we will be able to show that independence implies multiplicativity, it is possible to construct examples where multiplicativity holds without independence.

Definition 9.34. Two random variables f and g are said to be **independent** if, for all real α and β, the events $(f = \alpha)$ and $(g = \beta)$ are independent. So the equation

$$\mathbb{P}(f = \alpha \text{ and } g = \beta) = \mathbb{P}(f = \alpha) \cdot \mathbb{P}(g = \beta)$$

would hold for all real α and β.

There are generalizations of the concept of independent random variables to more than two random variables, but we will not cover them.

Lemma 9.35. If $h : \Omega \to \mathbb{R}$ is a random variable and $h(\Omega)$ denotes its range, then

$$\mathbb{E}(h) = \sum_{z \in h(\Omega)} z\mathbb{P}(h = z).$$

Proof. Let $h(\Omega) = \{u_1, \ldots, u_k\}$. The preimage principle (Theorem 2.1) asserts that

$$\langle h^{-1}(u_1), \ldots, h^{-1}(u_k)\rangle$$

is a generalized partition of Ω. Since $h(\Omega)$ is the range and not merely a codomain of h, each $h^{-1}(u_i)$ is non-empty. This means the set

$$P = \{h^{-1}(u_1), \ldots, h^{-1}(u_k)\}$$

is a partition of Ω. Then the expected value of h can be written as

$$\begin{aligned}\mathbb{E}(h) &= \sum_{\omega\in\Omega} p(\omega)h(\omega) = \sum_{i=1}^{k} \sum_{\omega\in h^{-1}(u_i)} p(\omega)h(\omega) \\ &= \sum_{i=1}^{k}\left(u_i \cdot \sum_{\omega\in(h=u_i)} p(\omega)\right) = \sum_{i=1}^{k} u_i\mathbb{P}(h = u_i) = \sum_{z\in h(\Omega)} z\mathbb{P}(h = z).\end{aligned}$$

□

Theorem 9.36. Let f and g be independent random variables with the common domain Ω. Then

$$\mathbb{E}(fg) = \mathbb{E}(f)\cdot\mathbb{E}(g).$$

Proof. Since Ω is a non-empty finite set, so are $f(\Omega)$ and $g(\Omega)$. Let M be the multiset of all sets of the form

$$S(x, y) = \{\omega \in \Omega : f(\omega) = x \text{ and } g(\omega) = y\},$$

where the multiset contains an element for each $(x, y) \in f(\Omega) \times g(\Omega)$. By independent multiplication (Theorem 3.1), M has

$$|f(\Omega)|\cdot|g(\Omega)|$$

elements, some of which might be empty. For each atomic event $\omega \in \Omega$, it is true that $\omega \in S(f(\omega), g(\omega))$, so each element of Ω belongs to some element of M. Moreover, if $(x, y) \neq (x', y')$ (for those unfamiliar with this notation, it means that $x \neq x'$ or $y \neq y'$, and possibly, but not necessarily, both), then

$$S(x, y) \cap S(x', y') = \varnothing.$$

So, M is a generalized partition of Ω. This allows us to compute

$$\begin{aligned}\mathbb{E}(fg) &= \sum_{\omega\in\Omega} p(\omega)f(\omega)g(\omega) = \sum_{(x,y)\in f(\Omega)\times g(\Omega)} \sum_{\omega\in S(x,y)} p(\omega)f(\omega)g(\omega) = \sum_{x\in f(\Omega)} \sum_{y\in g(\Omega)} \sum_{\omega\in S(x,y)} xy\cdot p(\omega) \\ &= \sum_{x\in f(\Omega)} \sum_{y\in g(\Omega)} \left(xy\cdot \sum_{\omega\in S(x,y)} p(\omega)\right) = \sum_{x\in f(\Omega)} \sum_{y\in g(\Omega)} xy\cdot\mathbb{P}(f = x \text{ and } g = y).\end{aligned}$$

By the independence of f and g, we get the equivalent expression

$$\sum_{x\in f(\Omega)} \sum_{y\in g(\Omega)} xy\cdot\mathbb{P}(f = x)\cdot\mathbb{P}(g = y) = \left(\sum_{x\in f(\Omega)} x\mathbb{P}(f = x)\right)\cdot\left(\sum_{y\in g(\Omega)} y\mathbb{P}(g = y)\right).$$

By Lemma 9.35, this is equal to $\mathbb{E}(f)\cdot\mathbb{E}(g)$ and we are done. □

An application of expected value that can be considered to be elementary is a non-constructive technique of proving existence, called the "probabilistic method," which we will now display.

Theorem 9.37 (Erdős' lower bound on Ramsey numbers)**.** If n and k are positive integers such that $n \leq 2^{\frac{k}{2}}$ and $k \geq 3$, then it is possible to colour the edges of a complete graph on n vertices with two colours such that there does not exist a monochromatic k-clique. As a result, the Ramsey number $R(k, k)$ satisfies the exponential lower bound

$$R(k, k) > 2^{\frac{k}{2}}.$$

Proof. Suppose n and k are as stated. Our technique will be to show that the expected value of the number X of monochromatic k-cliques is strictly less than 1. As a result, it will turn out that we can force the existence of a colouring with no monochromatic k-clique.
Let the list of k-cliques in the complete n-graph K_n be

$$C_1, C_2, \ldots, C_{\binom{n}{k}},$$

since there are $\binom{n}{k}$ ways to select exactly k vertices out of n vertices. For each $i \in \left[\binom{n}{k}\right]$, let X_i be the random variable defined on the domain of colourings of K_n (we will say that S is a colouring) such that

$$X_i(S) = \begin{cases} 1 & \text{if } C_i \text{ is monochromatic in } S \\ 0 & \text{otherwise} \end{cases}.$$

By the linearity of expectation, the expected value of the number of monochromatic k-cliques is

$$\mathbb{E}[X] = \mathbb{E}\left[\sum_{i=1}^{\binom{n}{k}} X_i\right] = \sum_{i=1}^{\binom{n}{k}} \mathbb{E}[X_i].$$

The only non-zero value of $\mathbb{E}[X_i]$ is 1, and it occurs if and only if C_i is monochromatic in a colouring. The probability of C_i being monochromatic in a random colouring is computed using the fact that the probability of each edge being the same colour is $\left(\frac{1}{2}\right)^{\binom{k}{2}}$ and there are 2 colours, yielding $2 \cdot \left(\frac{1}{2}\right)^{\binom{k}{2}}$. Thus,

$$\mathbb{E}[X] = \sum_{i=1}^{\binom{n}{k}} \mathbb{E}[X_i] = \sum_{i=1}^{\binom{n}{k}} 2 \cdot \left(\frac{1}{2}\right)^{\binom{k}{2}} = \binom{n}{k} \cdot 2^{1-\binom{k}{2}}.$$

Now we claim that

$$\mathbb{E}[X] = \binom{n}{k} \cdot 2^{1-\binom{k}{2}} < 1.$$

Using $n \leq 2^{\frac{k}{2}}$, an initial upper bound is

$$\binom{n}{k} \cdot 2^{1-\binom{k}{2}} = \frac{n(n-1)\cdots(n-k+1)}{k!} \cdot 2^{1-\binom{k}{2}} < \frac{n^k}{k!} \cdot 2^{1-\binom{k}{2}} \leq \frac{(2^{\frac{k}{2}})^k}{k!} \cdot 2^{1-\frac{k(k-1)}{2}} = \frac{2^{\frac{k}{2}+1}}{k!}.$$

We wish to prove that this is strictly less than 1, so we will prove that

$$2^{\frac{k}{2}+1} < k!$$

by induction on $k \geq 3$. Working backwards in the base case, $k = 3$, yields

$$2^{\frac{3}{2}+1} < 3! \iff 2^{\frac{5}{2}} < 6 \iff 32 < 36,$$

which is true. For the induction hypothesis, suppose there exists a $k \geq 3$ such that $2^{\frac{k}{2}+1} < k!$. Then

$$2^{\frac{k+1}{2}+1} = 2^{\frac{k}{2}+1} \cdot \sqrt{2} < k! \cdot \sqrt{2} < k! \cdot (k+1) = (k+1)!,$$

as desired. By Markov's inequality (Theorem 9.33),

$$\mathbb{P}(X \geq 1) \leq \frac{\mathbb{E}[X]}{1} < 1,$$
$$0 < 1 - \mathbb{P}(X \geq 1) = \mathbb{P}(X = 0),$$

since there are no values that X can take on strictly between 0 and 1. As the probability of there being no monochromatic k-clique is strictly positive, it means that there exists a colouring of K_n in which there is no monochromatic k-clique. □

There is much more to be said about finite probability spaces than we have shown, let alone for countably infinite discrete or general probability spaces. Some further topics are conditional expectation, variance and moments, the weak law of large numbers, and various inequalities involving expectation. The reader is encouraged to learn about these topics from other sources.

Chapter 10

Classic Recursions

"There is no clear-cut distinction between example and theory."

– *Michael Atiyah*

"A heavy warning used to be given [by lecturers] that pictures are not rigorous; this has never had its bluff called and has permanently frightened its victims into playing for safety. Some pictures, of course, are not rigorous, but I should say most are (and I use them whenever possible myself)."

– *John Edensor Littlewood, A Mathematician's Miscellany*

Recursive counting allows us to determine the cardinalities of finite sets in a sequence by expressing the cardinalities of sets of sufficiently large index in terms of the cardinalities of sets that came earlier in the sequence. In a sense, it is about determining the future in terms of the past. Two famous examples of recursively defined sequences with combinatorial interpretations that we will study are the Fibonacci numbers and the Catalan numbers.

10.1 Recursive Counting

"Enumerative combinatorics" is about determining the cardinality of every set in a sequence of finite sets. So, it involves solving countably many problems at once, as opposed to one concrete problem with one numerical answer. We have seen simple examples of enumerative problems, such as determining the number of bijections $f : [n] \to [n]$ (Problem 3.14) and computing the cardinality of the power set of $\mathcal{P}([n])$ of $[n]$ for each positive integer n (Example 3.2). Sometimes, it is possible to answer such questions by establishing a recurrence relation, which is an equation that expresses all terms of sufficiently high index in terms of terms of lower index, and then solving the recurrence relation as if it is an algebraic or inductive problem. This technique also requires manually computing a few initial terms. Let us see some simple problems that can be solved in this way.

Example 10.1. For each positive integer n, recall from the solution to Example 3.2 that $\{0,1\}^n$ denotes the set of n-tuples of 0's and 1's. Show that

$$|\{0,1\}^{k+1}| = 2 \cdot |\{0,1\}^k|$$

for each positive integer k and use this to find a general formula for $|\{0,1\}^n|$ for positive integers n.

Solution. Let k be a positive integer. We map each element of $\{0,1\}^{k+1}$ to an element of $\{0,1\}^k$ by shearing off the rightmost element. This is a 2-to-1 correspondence, so the correspondence principle (Theorem 3.19) gives the recursive relation

$$|\{0,1\}^{k+1}| = 2 \cdot |\{0,1\}^k|.$$

Note that

$$|\{0,1\}^1| = |\{0,1\}| = 2.$$

Since we start with 2 and multiply by 2 every time the exponent increases by 1, the answer is that

$$|\{0,1\}|^n = 2^n$$

for every positive integer n. Formally, we would prove this by induction: we did a manual computation for the base case $n = 1$, and we can use the recursive relation in the inductive step. □

Example 10.2. In this example, we will pretend like we do not know that Pascal's triangle consists of binomial coefficients. All we know is that we have a triangular array of numbers such that there is a 1 on the far left and a 1 on the far right of each row and each other number is the sum of the two numbers directly to its top-left and top-right.

$$\begin{array}{ccccccccccc}
 & & & & & 1 & & & & & \\
 & & & & 1 & & 1 & & & & \\
 & & & 1 & & 2 & & 1 & & & \\
 & & 1 & & 3 & & 3 & & 1 & & \\
 & 1 & & 4 & & 6 & & 4 & & 1 & \\
\iddots & & & & & \vdots & & & & & \ddots
\end{array}$$

If the top row is called row 0, then, for each non-negative integer n, determine the sum of the numbers in row n.

Solution. For each non-negative integer n, let S_n be the sum of the numbers in row n. The key observation needed to form a recursive relation for the S_n is that the sum of each row below row 0 consists of exactly two copies of each element of the previous row. For example,

$$\begin{aligned}
S_4 &= 1 + 4 + 6 + 4 + 1 \\
&= 1 + (1 + 3) + (3 + 3) + (3 + 1) + 1 \\
&= 2 \cdot (1 + 3 + 3 + 1).
\end{aligned}$$

In this way, $S_{k+1} = 2 \cdot S_k$ for each non-negative integer k. Since $S_0 = 1$, it follows from the same induction argument as in Example 10.1 that $S_n = 2^n$ for all non-negative integers n. □

Problem 10.3. There is a tetrahedral stack of identical spheres that descends downward infinitely, where each horizontal level consists of an equilateral triangle of spheres. The sphere at the top (level 0) is assigned the number 1. At every level below level 0, the number assigned to each sphere is the sum of the numbers assigned to the spheres that are tangent to this sphere in the level directly above. Here are levels 0 through 3:

$$1 \qquad \begin{array}{ccc} & 1 & \\ 1 & & 1 \end{array} \qquad \begin{array}{ccccc} & & 1 & & \\ & 2 & & 2 & \\ 1 & & 2 & & 1 \end{array} \qquad \begin{array}{ccccccc} & & & 1 & & & \\ & & 3 & & 3 & & \\ & 3 & & 6 & & 3 & \\ 1 & & 3 & & 3 & & 1 \end{array}$$

For each non-negative integer n, determine the sum of the numbers assigned to the spheres in level n.

Problem 10.4. Recall from Definition 7.22 that, for each positive integer n, the nth Bell number B_n is the number of set partitions of $[n]$. We define $B_0 = 1$. For each integer $n \geq 1$, prove that

$$B_{n+1} = \sum_{k=0}^{n} \binom{n}{k} B_k.$$

10.2 Fibonacci Numbers

We saw an application of the Fibonacci numbers in Example 4.10. Let us study these numbers in greater detail now.

Definition 10.5. The **Fibonacci numbers** $(F_n)_{n=0}^{\infty}$ are defined by

$$F_n = \begin{cases} 0 & \text{if } n = 0 \\ 1 & \text{if } n = 1 \\ F_{n-1} + F_{n-2} & \text{if } n \geq 2 \end{cases}.$$

The first few Fibonacci numbers are

$$0, 1, 1, 2, 3, 5, 8, 13, 21, \ldots.$$

Example 10.6 (Cassini's identity)**.** For every positive integer n,

$$F_{n+1}F_{n-1} - F_n^2 = (-1)^n.$$

Solution. This solution uses 2×2 matrix multiplication and determinants, as described in Appendix A. We claim that, for every positive integer n,

$$\begin{pmatrix} 1 & 1 \\ 1 & 0 \end{pmatrix}^n = \begin{pmatrix} F_{n+1} & F_n \\ F_n & F_{n-1} \end{pmatrix}.$$

This can be proven by induction as follows. The assertion can be immediately verified to be true for $n = 1$. Assuming that it holds for some positive integer n, the inductive step is

$$\begin{aligned} \begin{pmatrix} 1 & 1 \\ 1 & 0 \end{pmatrix}^{n+1} &= \begin{pmatrix} 1 & 1 \\ 1 & 0 \end{pmatrix}^n \cdot \begin{pmatrix} 1 & 1 \\ 1 & 0 \end{pmatrix} = \begin{pmatrix} F_{n+1} & F_n \\ F_n & F_{n-1} \end{pmatrix} \cdot \begin{pmatrix} 1 & 1 \\ 1 & 0 \end{pmatrix} \\ &= \begin{pmatrix} F_{n+1} + F_n & F_{n+1} \\ F_n + F_{n-1} & F_n \end{pmatrix} = \begin{pmatrix} F_{n+2} & F_{n+1} \\ F_{n+1} & F_n \end{pmatrix}. \end{aligned}$$

By the multiplicativity of the determinant, we get

$$\begin{aligned} (-1)^n &= \left[\det \begin{pmatrix} 1 & 1 \\ 1 & 0 \end{pmatrix}\right]^n = \det\left[\begin{pmatrix} 1 & 1 \\ 1 & 0 \end{pmatrix}^n\right] \\ &= \det \begin{pmatrix} F_{n+1} & F_n \\ F_n & F_{n-1} \end{pmatrix} = F_{n+1}F_{n-1} - F_n^2 \end{aligned}$$

for every positive integer n. □

The Fibonacci numbers satisfy a plethora of such identities. For example, there exists a generalization of Cassini's identity called Catalan's identity (unrelated to the Catalan numbers) and an even more general identity called Vadja's identity. Discussing these would take us too far into the world of obscure identities. We will not dwell on Fibonacci identities beyond the basic ones listed in the next problem.

Problem 10.7. Prove the following identities for all positive integers n.

1. $\displaystyle\sum_{k=1}^{n} F_k = F_{n+2} - 1$
2. $\displaystyle\sum_{k=0}^{n-1} F_{2k+1} = F_{2n}$
3. $\displaystyle\sum_{k=1}^{n} F_{2k} = F_{2n+1} - 1$
4. $\displaystyle\sum_{k=1}^{n} F_k^2 = F_nF_{n+1}$

In the context of double counting, we have seen that, for each integer $n \geq 3$, F_n is the number of $(n-2)$-tuples of 0's and 1's such that no two adjacent symbols are 1's. This interpretation led to a formula for F_n in terms of binomial coefficients (Example 4.10), but the formula was not closed because the number of terms in it increases as n increases. Now we will see a bizarre closed formula for F_n. It is strange enough that, at first sight, one might suspect that it does not even produce integer values.

Theorem 10.8 (Binet's formula)**.** Let $\phi = \dfrac{1+\sqrt{5}}{2}$, which is called the **golden ratio**, and let $\psi = \dfrac{1-\sqrt{5}}{2}$, which is the radical conjugate of the golden ratio. Then

$$F_n = \frac{\phi^n - \psi^n}{\sqrt{5}}$$

for every non-negative integer n.

Proof. As a disclosure, we will use an infinite generating function to find the formula. Since the generating function will have infinitely many terms, we will be using results about convergence. This would normally be troublesome without some background in analysis, but luckily our work on infinite geometric series will translate to this scenario nicely. The real issue is that we will be assuming is that two generating functions (interpreted as functions in the variable x in some interval around 0) are equal if and only if they have the same coefficient for each pair of corresponding terms. We will not justify this as usual, because the standard proof relies on the derivative. Also, in the summations below, whenever the expression 0^0 occurs, it is just convenient notation to denote the number 1; this is not a license to universally claim that $0^0 = 1$, as the expression is ordinarily undefined.
By the Fibonacci recursive relation, the Fibonacci generating function can be defined and manipulated as

$$\begin{aligned} F(x) &= \sum_{k=0}^{\infty} F_k x^k = 0 + x + \sum_{k=2}^{\infty} F_k x^k = x + \sum_{k=0}^{\infty} F_{k+2} x^{k+2} \\ &= x + \sum_{k=0}^{\infty} (F_{k+1} + F_k) x^{k+2} \\ &= x + x^2 \cdot \sum_{k=0}^{\infty} F_k x^k + x \cdot \sum_{k=0}^{\infty} F_{k+1} x^{k+1} \\ &= x + x^2 \cdot \sum_{k=0}^{\infty} F_k x^k + x \cdot \left(F_0 + \sum_{k=1}^{\infty} F_k x^k \right) \\ &= x + x^2 \cdot F(x) + x \cdot F(x). \end{aligned}$$

Isolating $F(x)$ in this equation yields

$$F(x) = \frac{x}{1 - x - x^2}.$$

By partial fraction decomposition, this can be written as

$$F(x) = \frac{1}{\sqrt{5}} \left(\frac{\phi_1}{\phi_1 - x} - \frac{\phi_2}{\phi_2 - x} \right),$$

where $\phi_1 = \dfrac{-1 + \sqrt{5}}{2}$ and $\phi_2 = \dfrac{-1 - \sqrt{5}}{2}$ are the roots of $x^2 + x - 1$.
For every real number r such that $|r| < 1$, we know that

$$\frac{1}{1 - r} = \sum_{k=0}^{\infty} r^k.$$

This formula for an infinite geometric series allows us to rewrite $F(x)$ as

$$\begin{aligned} F(x) &= \frac{1}{\sqrt{5}} \left(\frac{1}{1 - \frac{x}{\phi_1}} - \frac{1}{1 - \frac{x}{\phi_2}} \right) \\ &= \frac{1}{\sqrt{5}} \sum_{k=0}^{\infty} \left(\left(\frac{x}{\phi_1} \right)^k \right) - \left(\frac{x}{\phi_2} \right)^k \\ &= \frac{1}{\sqrt{5}} \sum_{k=0}^{\infty} \left(\frac{1}{\phi_1^k} - \frac{1}{\phi_2^k} \right) x^k, \end{aligned}$$

where the convergence holds for all real x such that $\left| \dfrac{x}{\phi_1} \right| < 1$ and $\left| \dfrac{x}{\phi_2} \right| < 1$. For such x,

$$\sum_{k=0}^{\infty} F_k x^k = F(x) = \sum_{k=0}^{\infty} \frac{1}{\sqrt{5}} \left(\frac{1}{\phi_1^k} - \frac{1}{\phi_2^k} \right) x^k.$$

By comparing coefficients of x^k and using the fact that $\phi_1\phi_2 = -1$ (this follows from Vieta's formulas for quadratics from Volume 1), we get that, for every non-negative integer n,

$$\begin{aligned} F_n &= \frac{1}{\sqrt{5}}\left(\frac{1}{\phi_1^n} - \frac{1}{\phi_2^n}\right) \\ &= \frac{1}{\sqrt{5}}((-\phi_2)^n - (-\phi_1)^n) \\ &= \frac{1}{\sqrt{5}}\left[\left(\frac{1+\sqrt{5}}{2}\right)^n - \left(\frac{1-\sqrt{5}}{2}\right)^n\right] \\ &= \frac{\phi^n - \psi^n}{\sqrt{5}}. \end{aligned}$$

□

Though we have seen it stated in some places that, in order to find formulas for other recursive sequences, it is possible to generalize this method of using generating functions in conjunction with partial fraction decomposition, we have generally not found this approach to be clean. We need a method that easily generalizes. Accordingly, the theory of linear recurrences will be the topic of Chapter 11.

Problem 10.9. For some positive integer n, there are n tiles $t_1, t_2, \ldots, t_n$ in a row in that order. You are playing a game where you are standing on t_1 and you wish to end up on t_n. From each tile t_i, you may jump to t_{i+1} or t_{i+2}. In how many ways can you reach t_n (meaning, you want to stand on t_n) from t_1?

10.3 Catalan Numbers

The renowned combinatorialist, Richard Stanley, has described the Catalan numbers as "probably the most ubiquitous sequence of numbers in mathematics." This is supported by the evidence that the longest entry in the On-Line Encyclopedia of Integer Sequences (OEIS) is that of the Catalan numbers [7].

Definition 10.10. The **Catalan numbers** $(C_n)_{n=0}^{\infty}$ are defined by $C_0 = 1$ and

$$C_{n+1} = \sum_{k=0}^{n} C_k C_{n-k}$$

for all non-negative integers n. The first few Catalan numbers are

$$1, 1, 2, 5, 14, 42, 132, 429, 1430, \ldots.$$

Definition 10.11. If n is a non-negative integer, then an **initial segment** of an n-tuple is the m-tuple formed by the leftmost m entries of the n-tuple for some non-negative integer $m \leq n$. A **Dyck word** is a $2n$-tuple for any non-negative integer n such that n of the entries are the symbol A and n of the symbols are B, and no initial segment of the $2n$-tuple contains strictly more B's than A's, though ties are allowed. For each non-negative integer n, let D_n denote the number the Dyck words of length $2n$.

Example. There are at least two ways of interpreting the meaning of Dyck words:

1. One interpretation is that, if the A's are replaced by open parenthesis (and the B's are replaced by closed parenthesis), then Dyck words produce precisely the ways of matching n open parentheses and n closed parentheses. For example,

$$AAABABBBAABB = ((()()))(()).$$

2. Dyck words may be visualized as "mountain ranges" on the Cartesian plane. Iterating through the symbols of a Dyck word from left to right, we start at $(0,0)$, and move from (p,q) to $(p+1,q+1)$ every time we hit an A and from (p,q) to $(p+1,q-1)$ every time we hit a B. Then Dyck words

produce exactly the mountain ranges where the final point lies on the x-axis and none of the points dip below the x-axis. For example

$$AAABABBBAABB$$

is given by

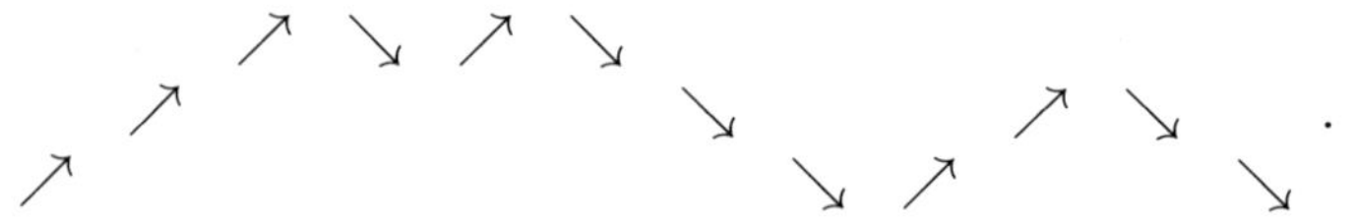

Note that, in this interpretation, each Dyck word results in a mountain range that has one more point than the number of symbols in the Dyck word, though the number of arrows in the mountain range is equal to the number of symbols in the Dyck word.

To be clear, the empty tuple is a Dyck word of length 0.

Lemma 10.12. For every positive integer n and every Dyck word d of length $2n$, there exist unique (possibly empty) Dyck words d_1 and d_2 such that

$$d = Ad_1Bd_2.$$

Proof. The mountain range visualization will be useful for conceptualizing this proof. We will say that a "tie" occurs at a certain index if the number of A's and B's from the first index up to and including that index are equal. For example $AAABABBBAABB$ has ties at indices 8 and 12. Let d be a Dyck word of length $2n$. The leftmost symbol of d must be A, otherwise the first initial segment of d will have more B's than A's. As for the location of the B in the decomposition Ad_1Bd_2, we claim that it must be the first index at which a tie occurs; this index must exist by the well-ordering principle because the set of such indices is non-empty due to a tie occurring at the final index.
It is not difficult to see that the tuple d_1 that is strictly in between these specific A and B is a Dyck word, as is the tuple d_2 directly to the right of this B. What remains to be shown is the uniqueness of the decomposition Ad_1Bd_2. As we have already stated, the leftmost symbol is always A, so we just have to prove the uniqueness of the location of the B. The potential indices of B are restricted to those indices at which the A's and B's are tied, otherwise d_1 will not have an equal number of A's and B's. Moreover, the index of B cannot be at any tie that is strictly to the right of the leftmost tie because then the number of B's will exceed the number of A's in d_1 at the index where the leftmost tie occurs in d. Thus, a unique decomposition Ad_1Bd_2 exists. □

Corollary 10.13. For every non-negative integer n,

$$D_{n+1} = \sum_{k=0}^{n} D_k D_{n-k}.$$

Since $D_0 = C_0$, this means that $D_n = C_n$ for all non-negative integers n, where $(D_n)_{n=0}^{\infty}$ are the Dyck numbers and $(C_n)_{n=0}^{\infty}$ is the Catalan sequence.

Proof. Let n be a non-negative integer n and d be a Dyck word of length $2(n+1) = 2n+2$. By Lemma 10.12, there exist unique Dyck words d_1 and d_2 such that $d = Ad_1Bd_2$. Since the length of d is $2n+2$, the combined lengths of d_1 and d_2 is n. The map that sends d to (d_1, d_2) is a bijection from the set of Dyck words of length $2n+2$ to the set of ordered pairs of Dyck words whose lengths sum to n, due to the uniqueness of the decomposition. The length of d_1 can be $k = 0, 1, 2, \ldots, n$ and the respective lengths of d_2 are $n, n-1, n-2, \ldots, 0$. By the bijection principle (Theorem 1.10), casework (Theorem 1.31), and independent multiplication (Theorem 3.1),

$$D_{n+1} = \sum_{k=0}^{n} D_k D_{n-k}.$$

Since this recursive relation is identical to the Catalan recursion and $D_0 = 1 = C_0$ (the only Dyck word of length 0 is the empty tuple), it can be proven by strong induction that $D_n = C_n$ for all non-negative integers n. □

Theorem 10.14 (Catalan numbers formula)**.** For every non-negative integer n,

$$C_n = \frac{1}{n+1}\binom{2n}{n}.$$

Proof. Suppose a and b are non-negative integers. If $a > b$, then let $S_>(a,b)$ denote the set of $(a+b)$-tuples of a copies of A and b copies of B such that every initial segment of each tuple has strictly more A's than B's. If $a \geq b$, then let $S_\geq(a,b)$ denote the set of $(a+b)$-tuples of a copies of A and b copies of B such that no initial segment of any tuple has strictly more B's than A's. Given non-negative integers a and b such that $a \geq b$, define the function

$$f : S_\geq(a,b) \to S_>(a+1,b)$$

by appending an A to the far left of each element of the domain. Then f is a surjection because the leftmost symbol of each element of $S_>(a+1,b)$ is always A and removing it yields an element of $S_\geq(a,b)$; f is also an injection because removing the leftmost A of an element of $S_>(a+1,b)$ yields a unique element of $S_\geq(a,b)$. Thus, we have a bijection. By the solution to Bertrand's ballot problem (Theorem 9.7),

$$\begin{aligned}
|S_\geq(a,b)| &= |S_>(a+1,b)| \\
&= \frac{a+1-b}{a+1+b}\cdot\binom{a+1+b}{b} \\
&= \frac{a+1-b}{a+1+b}\cdot\binom{a+1+b}{a+1} \\
&= \frac{a+1-b}{a+1+b}\cdot\frac{a+1+b}{a+1}\cdot\binom{a+b}{a} \\
&= \frac{a+1-b}{a+1}\cdot\binom{a+b}{b}.
\end{aligned}$$

For each non-negative integer n, taking $a = b = n$ yields

$$C_n = D_n = |S_\geq(n,n)| = \frac{1}{n+1}\binom{2n}{n}.$$

In summary, we used the discrete intermediate value theorem (Lemma 9.6) to solve Bertrand's ballot problem earlier. A moment ago, we showed that counting Dyck words is the same as computing Catalan numbers (Corollary 10.13). Finally, we used Bertrand's ballot problem to count Dyck words, resulting in a compact formula for Catalan numbers. ☐

The Catalan numbers turn out to be the solution to hundreds of seemingly disparate problems in enumerative combinatorics. Many of these problems have been collected by Richard Stanley in [8]. We will see two of the more well-known problems now, one from geometry and one related to graph theory.

Definition 10.15. A **triangulation** of a polygon P is a collection of triangles such that their interiors are non-intersecting and the union of the triangles is P. A triangulation of P is said to have **no extra vertices** if the vertices of the triangles in the triangulation are chosen from only the vertices of P.

Example 10.16. For each positive integer n, and for all convex $(n+2)$-gons P_{n+2}, show that the number of triangulations of P_{n+2} that have no extra vertices is the nth Catalan number C_n. Here, we consider the vertices of the polygon to be distinguishable when determining which triangulations are distinct from each other. Regarding the definition of convexity, the most useful one here is that the interiors of all diagonals lie in the interior of the polygon (see Volume 4).

Solution. Our proof is by induction on $n \geq 1$. For $n = 1$, the only triangulation of a triangle (since $P_{n+2} = P_{1+2} = P_3$) that has no extra vertices is the triangle itself. So every triangle has exactly $1 = C_1$ triangulation. Now suppose the result holds up to and including some positive integer n. Let P_{n+3} be a convex $(n+3)$-gon. The idea is that, if we fix a particular edge AB of P_{n+3}, then every triangulation of P_{n+3} contains exactly one triangle with AB as an edge of the triangle. Now we do casework on the $n+1$ possibilities for the third vertex of the triangle.

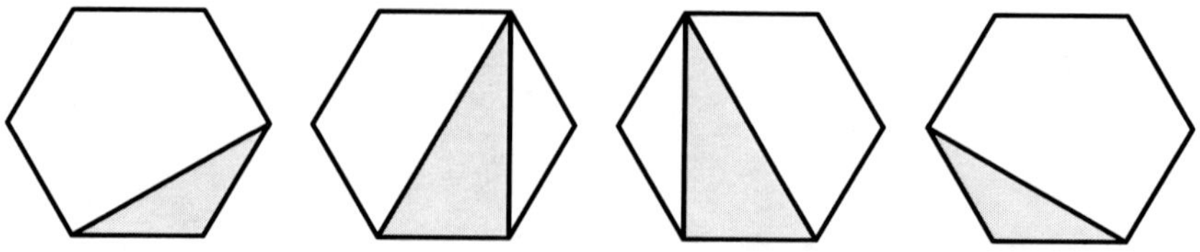

For $m = 3, 4, \ldots, n+2$, let N_m denote the number of triangulations of convex m-gons that have no extra vertices. By removing the triangle attached to AB, we can invoke the strong induction hypothesis on the two leftover pieces to get that the number of triangulations of P_{n+3} with no new vertices is

$$N_{n+2} + (N_3 N_{n+1} + N_4 N_n + \cdots + N_n N_4 + N_{n+1} N_3) + N_{n+2}.$$

Technically, we have used the fact that removing the triangle attached to AB results in one or two convex components. This is due to the fact that convex polygons are, by definition, formed by the intersection of half-planes (see Volume 4).
By defining $N_2 = 1 = C_0$ for the sake of convenient notation, we can write this sum more comfortably as

$$\sum_{k=2}^{n+2} N_k N_{n+4-k} = \sum_{k=0}^{n} N_{k+2} N_{n+2-k}.$$

For each integer $k = 0, 1, 2, \ldots, n$, the subscripts $k+2$ and $n+2-k$ lie in the interval $[2, n+2]$. So we can apply the strong induction hypothesis to rewrite the sum as

$$\sum_{k=0}^{n} C_k C_{n-k} = C_{n+1}.$$

Therefore, the result holds by strong induction. So, if $t \geq 3$ is a fixed integer, we may speak freely of "the" number of triangulations with no new vertices of any convex t-gon because the number is the same for all convex t-gons. □

Definition 10.17. A **rooted full binary tree** is constructed as follows: There is a special node called the **root**. Each node is descended from a unique node, with the former being called a **child** of the latter and the latter being called the **parent** of the former. Each node has no children or two distinguishable children that are called the **left child** and the **right child**. The adjective "rooted" comes from the fact that there is a special node called the root, and the adjective "full" comes from the fact that a node cannot have just one child.

Example. The following are the rooted full binary trees on $n = 1, 2, 3$ parent nodes. The top row is for $n = 1$, the second row is for $n = 2$, and the third and fourth rows are for $n = 3$.

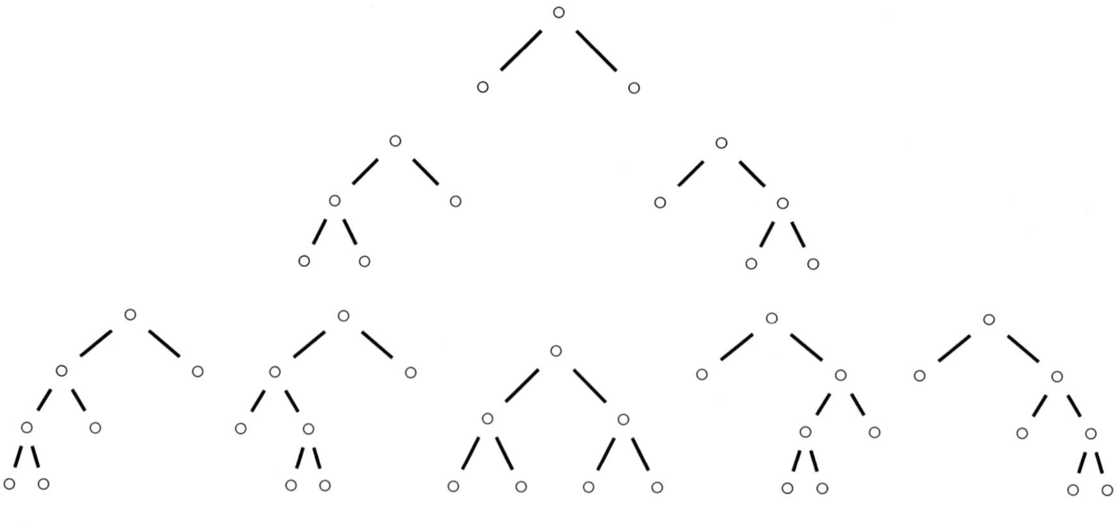

For $n = 0$, there is only one possibility which consists of a single node.

Problem 10.18. For each non-negative integer n, determine the number of rooted binary trees with exactly n nodes that are parents of some nodes.

Chapter 11

Linear Recurrences

"I couldn't do it. I couldn't reduce it to the freshman level. That means we really don't understand it."

– *Richard P. Feynman*

Having seen the merit of using recursive relations to solve combinatorial problems in Chapter 10, we are led to studying sequences of numbers satisfying recursive relations as a purely algebraic matter without any attention to potential combinatorial contexts. While recursive relations can come in an infinite variety of forms, the most common ones are those that satisfy a sense of linearity. We will find closed formulas for special cases of such recurrences to some degree of generality.

11.1 Non-homogeneous

As we know, a recursive relation on a sequence is an equation that expresses terms of sufficiently high index in terms of terms of lower index. Note that it is an abuse of terminology, albeit a common one, to say that a *sequence* is recursive. This is because a sequence of numbers might satisfy a recursive relation as well as a closed formula, with the former holding no significance in a particular context. For example, the sequence of non-negative integers

$$(a_n)_{n=0}^{\infty} = (0, 1, 2, 3, \ldots)$$

satisfies the recursive relation $a_n = a_{n-1} + 1$ for all positive integers n, but this fact is not relevant in every setting (though it is used in the formal study of the rules of arithmetic, namely the Peano axioms). In summary, being recursive is property of a relation that is a particular presentation of a sequence; it is not inherent to the sequence itself. This is similar to how piecewise functions are not inherently piecewise, but rather piecewise-defined.

Definition 11.1. If no sequence is specified, but a recurrence relation is stated to hold, starting at a certain index, then the relation is said to be **unrestricted**. Any sequence for which the relation holds for all sufficiently large indices is said to **satisfy** or **solve** or be a **solution** to the recurrence relation.

Definition 11.2. If $(z_n)_{n=0}^{\infty}$ is a sequence of complex numbers and k is a positive integer, then a **linear recurrence relation** of **depth** k on the sequence is an equation

$$z_n = p(n) + \sum_{i=1}^{k} c_i z_{n-i}$$

that is satisfied for all integers $n \geq k$. Here, p is a polynomial with complex coefficients, and $c_1, c_2, \ldots, c_k$ are complex constants such that $c_k \neq 0$ (if c_k were zero, then the depth would be less than k). If p is the zero polynomial, then the linear recurrence relation is said to be **homogeneous**; otherwise, it is **non-homogeneous** and the degree of p is said to be the **degree** of the linear recurrence relation.

Example. The Fibonacci recurrence relation is linear and homogeneous, whereas the Catalan recurrence relation is not even linear.

Problem 11.3. Suppose we have an unrestricted linear recurrence relation

$$z_n = p(n) + \sum_{i=1}^{k} c_i z_{n-i}$$

of depth k. Show that for any specific values of $z_0, z_1, \ldots, z_{k-1}$, there is exactly one possible sequence $(z_n)_{n=0}^{\infty}$ that satisfies this recurrence relation. For example, this means that there is exactly one possible sequence that starts with 0 and 1 and satisfies the Fibonacci recurrence relation.

Problem 11.3 allows us to make the following definition.

Definition 11.4. Given an unrestricted linear recurrence relation of depth k, if we fix the first k numbers $z_0, z_1, \ldots, z_{k-1}$ among the set of sequences that satisfy the recurrence relation, then this list of k numbers is said to be the **initial conditions**. As we saw in Problem 11.3, having initial conditions allows us to speak of "the" unique sequence that satisfies the initial conditions, and also satisfies the recurrence relation for indices greater than or equal to the depth k.

Momentarily, we will see that solving a non-homogeneous linear recurrence can always be boiled down to solving a homogeneous linear recurrence, using a technique called finite differences. This is a general philosophy in mathematics, where we wish to "reduce" what is complicated by showing that it suffices to solve what is simple. The following problem introduces one stand-alone reduction technique for linear recurrences, though it does not always work.

Problem 11.5. Consider the example of the recurrence relation

$$\alpha_n = 1 + \alpha_{n-1} + \alpha_{n-2}$$

with fixed values of α_0 and α_1. By adding 1 to both sides, we get

$$1 + \alpha_n = (1 + \alpha_{n-1}) + (1 + \alpha_{n-2}),$$

so it suffices to solve for

$$(\beta_n)_{n=0}^{\infty} = (1 + \alpha_n)_{n=0}^{\infty}$$

instead. That is, $\beta_n = \beta_{n-1} + \beta_{n-2}$ with initial conditions $\beta_0 = 1 + \alpha_0$ and $\beta_1 = 1 + \alpha_1$. Show that the linear recurrence

$$a_n = c + \sum_{i=1}^{k} c_i a_{n-i}$$

of depth k, where c is a non-zero constant that is the polynomial term of the recurrence, may be reduced to a homogeneous linear recurrence of depth k by adding a constant to both sides if and only if

$$c_1 + c_2 + \cdots + c_k \neq 1.$$

While the technique in Problem 11.5 works well in most cases where the polynomial term is a non-zero constant, we need a reduction method that works in general, as follows.

Theorem 11.6. Given a non-homogeneous linear recurrence relation of depth k and degree d with given initial conditions, suppose $(a_n)_{n=0}^{\infty}$ is the unique sequence that satisfies the recurrence relation for all indices $n \geq k$ and agrees with the initial conditions. Then $(a_n)_{n=0}^{\infty}$ satisfies a homogeneous linear recurrence relation of depth at most $k + d + 1$ for all indices $n \geq k + d + 1$.
Note that, while the original initial conditions prescribe the first k numbers in the sequence, if new depth of at most $k + d + 1$ is strictly greater than the original depth k, then this is not a problem. We can use the original non-homogeneous recurrence relation to manually compute more terms in the sequence before making use of the new homogeneous recurrence relation.

Proof. We will prove this by strong induction on the degree $d \geq 0$. For each d, we will prove the assertion simultaneously for all depths $k \geq 1$. In the base case $d = 0$, suppose the recurrence relation

$$a_n = c + \sum_{i=1}^{k} c_i a_{n-i}$$

holds, where c is a non-zero constant and k is any depth. Then

$$a_{n+1} = c + \sum_{i=1}^{k} c_i a_{n+1-i}.$$

Subtracting the former equation from the latter and rearranging yields

$$a_{n+1} = (1 + c_1)a_n + \sum_{i=1}^{k-1} (c_{i+1} - c_i)a_{n-i} - c_k a_{n-k},$$

which is a homogeneous linear recurrence relation of depth

$$k + 1 = k + 0 + 1 = k + d + 1$$

that holds for all integers n such that the index satisfies $n + 1 \geq k + 1$. This establishes the base case.
Now, suppose the result holds for all non-negative degrees less than or equal to d. Suppose we have a recurrence relation

$$a_n = p(n) + \sum_{i=1}^{k} c_i a_{n-i},$$

where p is a polynomial of degree $d + 1$ and the depth k is any positive integer. Then subtracting this from

$$a_{n+1} = p(n + 1) + \sum_{i=1}^{k} c_i a_{n+1-i}$$

and rearranging the result yields the linear recurrence relation

$$a_{n+1} = (p(n + 1) - p(n)) + (1 + c_1)a_n + \sum_{i=1}^{k-1} (c_{i+1} - c_i)a_{n-i} - c_k a_{n-k},$$

which has depth $k + 1$, degree at most d (because $p(n + 1)$ and $p(n)$ have the same leading term that gets eliminated upon taking their difference), and holds for all integers n such that the index satisfies $n + 1 \geq k + 1$. By invoking the strong induction hypothesis, the unique solution $(a_n)_{n=0}^{\infty}$ to this recurrence satisfies a homogeneous linear recurrence of depth at most

$$(k + 1) + d + 1 = k + (d + 1) + 1$$

for all indices $n \geq k + d + 2$. This completes the induction.
This process of taking differences allows us to explicitly construct such a homogeneous linear recurrence relation. □

11.2 Homogeneous

In this section, we will be using the expression 0^0 to denote the number 1. This is not an accepted computation in arithmetic, but briefly assuming that this notation is acceptable will allow us to replace 1 with 0^0 in some patterns, such as sequences and summations, thereby making their presentation more uniform across the indices.

Problem 11.7. Suppose t is a positive integer and

$$(b_{1,n})_{n=0}^{\infty}, (b_{2,n})_{n=0}^{\infty}, \ldots, (b_{t,n})_{n=0}^{\infty}$$

are t sequences that each individually satisfy the unrestricted homogeneous linear recurrence relation

$$a_n = \sum_{i=1}^{k} c_i a_{n-i}$$

of depth k for all indices $n \geq k$. Prove that, for any constants $\beta_1, \beta_2, \ldots, \beta_t$, the sequence

$$(\gamma_n)_{n=0}^{\infty} = (\beta_1 b_{1,n} + \beta_2 b_{2,n} + \cdots + \beta_t b_{t,n})_{n=0}^{\infty}$$

also satisfies the unrestricted recurrence. This justifies the adjective "linear."

The fact that Binet's formula for the Fibonacci numbers (Theorem 10.8) involves the exponential growth of both of its terms makes us wonder if the solutions of other homogeneous linear recurrences have a similarly exponential nature. This leads to the following definitions.

Definition 11.8. A solution to an unrestricted homogeneous linear recurrence relation is said to be **exponential** if there exists a non-zero complex constant z such that the solution is equal to $(z^n)_{n=0}^{\infty}$. If the recurrence relation is $a_n = \sum_{i=1}^{k} c_i a_{n-i}$, then $(z^n)_{n=0}^{\infty}$ solves it if and only if $z^n = \sum_{i=1}^{k} c_i z^{n-i}$ for every integer $n \geq k$. Dividing both sides by z^{n-k}, which is a reversible step, yields $z^k = \sum_{i=1}^{k} c_i z^{k-i}$. Accordingly, we define the **characteristic polynomial** of the recurrence relation to be the function $f : \mathbb{C} \to \mathbb{C}$, defined by

$$f(x) = x^k - \sum_{i=1}^{k} c_i x^{k-i}.$$

Since the constant term c_k of the characteristic polynomial is non-zero (this is due to the depth of the recurrence relation being k), 0 cannot be a root. So $(z^n)_{n=0}^{\infty}$ is an exponential solution to an unrestricted homogeneous linear recurrence if and only if z is a root of its characteristic polynomial. Importantly, note that, by the fundamental theorem of algebra, the characteristic polynomial has k roots if we include each root as many times as its multiplicity.

Definition 11.9. If n is a positive integer, then an $n \times n$ **Vandermonde matrix** (unrelated to Vandermonde's identity) is an $n \times n$ matrix of the form

$$\begin{bmatrix} 1 & \alpha_1 & \alpha_1^2 & \cdots & \alpha_1^{n-1} \\ 1 & \alpha_2 & \alpha_2^2 & \cdots & \alpha_2^{n-1} \\ 1 & \alpha_3 & \alpha_3^2 & \cdots & \alpha_3^{n-1} \\ \vdots & \vdots & \vdots & \ddots & \vdots \\ 1 & \alpha_n & \alpha_n^2 & \cdots & \alpha_n^{n-1} \end{bmatrix},$$

where $\alpha_1, \alpha_2, \ldots, \alpha_n$ are complex numbers. So, the entry at the intersection of row i and column j is α_i^{j-1} (here, we used the aforementioned notation of $0^0 = 1$ if $\alpha_i = 0$ and $j = 1$). The determinant of this Vandermonde matrix is equal to

$$\prod_{1 \leq i < j \leq n} (\alpha_j - \alpha_i) = \prod_{i=1}^{n-1} \prod_{j=i+1}^{n} (\alpha_j - \alpha_i),$$

but we will not prove this because we have not introduced a precise definition of the general $n \times n$ determinant. Computation of this determinant is sometimes an exercise in a first course in linear algebra. Note that if $\alpha_1, \alpha_2, \ldots, \alpha_n$ are distinct, then the Vandermonde determinant is non-zero because each multiplicand $(\alpha_j - \alpha_i)$ is non-zero.

Theorem 11.10. Suppose k is a positive integer and that we have a homogeneous linear recurrence relation

$$a_n = \sum_{i=1}^{k} c_i a_{n-i}$$

of depth k with fixed values of $a_0, a_1, \ldots, a_{k-1}$ as initial conditions. If the k roots $z_1, z_2, \ldots, z_k$ of the characteristic polynomial are distinct, then there exist unique complex numbers $v_1, v_2, \ldots, v_k$ such that the solution

$$a_n = \sum_{j=1}^{k} v_j z_j^n$$

holds for all non-negative integers n.

Proof. First we will show that, if unique complex numbers

$$v_1, v_2, \ldots, v_k$$

exist such that the equation

$$a_n = \sum_{j=1}^{k} v_j z_j^n$$

holds for $n = 0, 1, 2, \ldots, k-1$, then the same equation holds for all non-negative integers n. After that, it will suffice to find unique v_i that satisfy the first k instances of this equation.
We will prove that the equation holds for all non-negative integers n by induction on $n \geq 0$. The base case is taken care of by the assumption for $n = 0, 1, 2, \ldots, k-1$. Now suppose the equation holds for the indices $0, 1, 2, \ldots, n-1$ for some integer n such that $n - 1 \geq k - 1$. Then the next number in the sequence is, by the recurrence relation, the strong induction hypothesis, and the discrete Fubini's principle,

$$\begin{aligned}
a_n &= \sum_{i=1}^{k} c_i a_{n-i} = \sum_{i=1}^{k} c_i \sum_{j=1}^{k} v_j z_j^{n-i} \\
&= \sum_{i=1}^{k} \sum_{j=1}^{k} c_i v_j z_j^{n-i} = \sum_{j=1}^{k} \sum_{i=1}^{k} c_i v_j z_j^{n-i} \\
&= \sum_{j=1}^{k} v_j \sum_{i=1}^{k} c_i z_j^{n-i} = \sum_{j=1}^{k} v_j z_j^n,
\end{aligned}$$

where, in the last step, we have used the fact that z_j is a root of the characteristic polynomial to get that $z_j^n = \sum_{i=1}^{k} c_i z_j^{n-i}$. This completes the induction.
So, it suffices to find unique complex numbers $v_1, v_2, \ldots, v_k$ such that

$$a_n = \sum_{j=1}^{k} v_j z_j^n$$

for $n = 0, 1, 2 \ldots, k-1$. This system of k equations can be written using matrix multiplication as

$$\begin{bmatrix} 1 & 1 & 1 & \cdots & 1 \\ z_1 & z_2 & z_3 & \cdots & z_k \\ z_1^2 & z_2^2 & z_3^2 & \cdots & z_k^2 \\ \vdots & \vdots & \vdots & \ddots & \vdots \\ z_1^{k-1} & z_2^{k-1} & z_3^{k-1} & \cdots & z_k^{k-1} \end{bmatrix} \begin{bmatrix} v_1 \\ v_2 \\ v_3 \\ \vdots \\ v_k \end{bmatrix} = \begin{bmatrix} a_0 \\ a_1 \\ a_2 \\ \vdots \\ a_{k-1} \end{bmatrix}$$

We will use some basic results from linear algebra now. The square matrix on the far left is the transpose (this means flipped across the diagonal from the top-left to the bottom-right) of a $k \times k$ Vandermonde matrix. Since determinants are preserved under applying the transpose, the determinant of this matrix is $\prod_{1 \leq i < j \leq k} (z_j - z_i)$, which is non-zero, due to the hypothesis that the roots of the characteristic polynomial are distinct. As such, the matrix has an inverse, and left-multiplying both sides of the matrix equation by it yields

$$\begin{bmatrix} v_1 \\ v_2 \\ v_3 \\ \vdots \\ v_k \end{bmatrix} = \begin{bmatrix} 1 & 1 & 1 & \cdots & 1 \\ z_1 & z_2 & z_3 & \cdots & z_k \\ z_1^2 & z_2^2 & z_3^2 & \cdots & z_k^2 \\ \vdots & \vdots & \vdots & \ddots & \vdots \\ z_1^{k-1} & z_2^{k-1} & z_3^{k-1} & \cdots & z_k^{k-1} \end{bmatrix}^{-1} \begin{bmatrix} a_0 \\ a_1 \\ a_2 \\ \vdots \\ a_{k-1} \end{bmatrix},$$

which shows that unique values of $v_1, v_2, \ldots, v_k$ exist. □

Problem 11.11. The **Lucas numbers** $(L_n)_{n=0}^{\infty}$ are defined by

$$L_n = \begin{cases} 2 & \text{if } n = 0 \\ 1 & \text{if } n = 1 \\ L_{n-1} + L_{n-2} & \text{if } n \geq 2 \end{cases}.$$

The first few Lucas numbers are

$$2, 1, 3, 4, 7, 11, 18, 29, 47, \ldots.$$

Find a closed formula for the Lucas numbers that holds for all non-negative indices n.

Problem 11.12. The **Pell numbers** $(P_n)_{n=0}^{\infty}$ are defined by

$$P_n = \begin{cases} 0 & \text{if } n = 0 \\ 1 & \text{if } n = 1 \\ 2P_{n-1} + P_{n-2} & \text{if } n \geq 2 \end{cases}.$$

The first few Pell numbers are

$$0, 1, 2, 5, 12, 29, 70, 169, 408, \ldots.$$

Find a closed formula for the Pell numbers that holds for all non-negative indices n.

Theorem 11.13. Suppose k is a positive integer and that we have a homogeneous linear recurrence relation

$$a_n = \sum_{i=1}^{k} c_i a_{n-i}$$

of depth k with fixed values of $a_0, a_1, \ldots, a_{k-1}$ as initial conditions. If the k roots $z_1, z_2, \ldots, z_k$ of the characteristic polynomial are all equal to a complex number w, then there exist unique complex numbers $u_0, u_1, \ldots, u_{k-1}$ such that the equation

$$a_n = (u_0 + u_1 n + u_2 n^2 + \cdots + u_{k-1} n^{k-1}) w^n$$

holds for all non-negative integers n. In other words, there exists a unique polynomial q with complex coefficients and of degree less than k such that $a_n = q(n) w^n$ for all non-negative integers n.

Proof. As in the case of distinct roots (Theorem 11.10), we will show that, if there exist unique complex numbers $u_0, u_1, \ldots, u_{k-1}$ such that

$$a_n = (u_0 + u_1 n + u_2 n^2 + \cdots + u_{k-1} n^{k-1}) w^n$$

for $n = 0, 1, 2, \ldots, k-1$, then this equation holds for all non-negative integers n. After that, it will suffice to find unique u_i that satisfy the first k instances of this equation.
The first observation is that, by binomial expansion, the characteristic polynomial equals

$$x^k - \sum_{i=1}^{k} c_i x^{k-i} = (x - w)^k = \sum_{i=0}^{k} (-1)^i \binom{k}{i} w^i x^{k-i}.$$

Cancelling x^k from both sides and rearranging the equation yields

$$\sum_{i=1}^{k} c_i x^{k-i} = \sum_{i=1}^{k} (-1)^{i+1} \binom{k}{i} w^i x^{k-i}.$$

By comparing coefficients, we find that $c_i = (-1)^{i+1} \binom{k}{i} w^i$ for $i = 1, 2, \ldots, k$.

Now we proceed by strong induction on $n \geq 0$. By assumption,

$$a_n = (u_0 + u_1 n + u_2 n^2 + \cdots + u_{k-1} n^{k-1}) w^n$$

for $n = 0, 1, 2, \ldots, k-1$, which takes care of the base case. As the induction hypothesis, suppose this equation holds for all indices $0, 1, 2, \ldots, n-1$ for some integer n such that $n-1 \geq k-1$. By the recurrence relation, the strong induction hypothesis, and the discrete Fubini's principle, the next number in the sequence is

$$\begin{aligned}
a_n &= \sum_{i=1}^{k} c_i a_{n-i} \\
&= \sum_{i=1}^{k} \left[\left((-1)^{i+1} \binom{k}{i} w^i \right) \left(w^{n-i} \sum_{j=0}^{k-1} u_j (n-i)^j \right) \right] \\
&= \sum_{i=1}^{k} \sum_{j=0}^{k-1} \left((-1)^{i+1} \binom{k}{i} w^i \right) \left(w^{n-i} u_j (n-i)^j \right) \\
&= w^n \cdot \sum_{i=1}^{k} \sum_{j=0}^{k-1} (-1)^{i+1} \binom{k}{i} u_j (n-i)^j \\
&= w^n \cdot \sum_{j=0}^{k-1} u_j \sum_{i=1}^{k} (-1)^{i+1} \binom{k}{i} (n-i)^j.
\end{aligned}$$

Above, wherever 0^0 occurs, it should be interpreted as 1. We need to evaluate the inner sum. If $j = 0$, then the inner sum is

$$\sum_{i=1}^{k} (-1)^{i+1} \binom{k}{i} = 1 - \sum_{i=0}^{k} (-1)^i \binom{k}{i} = 1 - (-1+1)^k = 1 = n^j.$$

If $j > 0$ then, according to Corollary 6.14, if k, n, j are positive integers such that $n \geq k > j$, then

$$\sum_{i=0}^{k} (-1)^i \binom{k}{i} (n-i)^j = 0.$$

Rearranging, we can evaluate the inner sum as

$$\sum_{i=1}^{k} (-1)^{i+1} \binom{k}{i} (n-i)^j = n^j.$$

Therefore, $a_n = w^n \cdot \sum_{j=0}^{k-1} u_j n^j$, which completes the induction.

Finally, we will show that there exists unique complex numbers $u_0, u_1, \ldots, u_{k-1}$ such that

$$a_n = (u_0 + u_1 n + u_2 n^2 + \cdots + u_{k-1} n^{k-1}) w^n$$

for $n = 0, 1, 2, \ldots, k-1$. This system of equations can be written in matrix form as

$$\begin{bmatrix} 1 & 0 & \cdots & 0 \\ w & w & \cdots & w \\ w^2 & 2w^2 & \cdots & 2^{k-1} w^2 \\ \vdots & \vdots & \ddots & \vdots \\ w^{k-1} & (k-1) w^{k-1} & \cdots & (k-1)^{k-1} w^{k-1} \end{bmatrix} \begin{bmatrix} u_1 \\ u_2 \\ u_3 \\ \vdots \\ u_k \end{bmatrix} = \begin{bmatrix} a_0 \\ a_1 \\ a_2 \\ \vdots \\ a_{k-1} \end{bmatrix}$$

By pulling powers of w out of rows (we can do this due to the scalar multiple property of determinants that is described in Theorem A.7), the determinant of the square matrix on the far left is

$$w^{\frac{k(k-1)}{2}} \cdot \det \begin{bmatrix} 1 & 0 & 0 & \cdots & 0 \\ 1 & 1 & 1 & \cdots & 1 \\ 1 & 2 & 2^2 & \cdots & 2^{k-1} \\ \vdots & \vdots & \vdots & \ddots & \vdots \\ 1 & (k-1) & (k-1)^2 & \cdots & (k-1)^{k-1} \end{bmatrix}.$$

This matrix is a $k \times k$ Vandermonde matrix with $\alpha_1 = 0, \alpha_2 = 1, \ldots, \alpha_k = k-1$. As these are distinct numbers, the determinant of the Vandermonde matrix is non-zero. Moreover, since the constant term c_k of the characteristic polynomial is not zero, the root w is non-zero. Therefore, the determinant of the original matrix is non-zero, making the matrix invertible, which means unique complex numbers $u_0, u_1, \ldots, u_{k-1}$ exist, similar to Theorem 11.10. □

Problem 11.14. In the proof of Theorem 11.13, the $k \times k$ Vandermonde matrix

$$\begin{bmatrix} 1 & 0 & 0 & \cdots & 0 \\ 1 & 1 & 1 & \cdots & 1 \\ 1 & 2 & 2^2 & \cdots & 2^{k-1} \\ \vdots & \vdots & \vdots & \ddots & \vdots \\ 1 & (k-1) & (k-1)^2 & \cdots & (k-1)^{k-1} \end{bmatrix}$$

came up. Use the stated formula for the Vandermonde determinant to show that the determinant of this matrix is $\prod_{i=1}^{k-1} i^{k-i}$ for all integers $k \geq 2$.

We have shown how to solve homogeneous linear recurrences when the characteristic polynomial has all distinct roots or all equal roots, but we have not addressed what occurs in between these extremes. It is possible to use linear algebra to solve the general problem, but the proof is beyond the scope of our exposition. For completeness, we state the theorem as follows, which can also be found in [1].

Theorem 11.15. Suppose k is a positive integer and that we have a homogeneous linear recurrence relation

$$a_n = \sum_{i=1}^{k} c_i a_{n-i}$$

of depth k with fixed values of $a_0, a_1, \ldots, a_{k-1}$ as initial conditions. Let the distinct roots of the characteristic polynomial be $z_1, z_2, \ldots, z_t$ for some positive integer t, with respective multiplicities $m_1, m_2, \ldots, m_t \geq 1$. Then there exist unique polynomials $p_1, p_2, \ldots, p_t$ of degrees strictly less than $m_1, m_2, \ldots, m_t$, respectively, such that

$$a_n = \sum_{i=1}^{t} p_i(n) \cdot z_i^n$$

for all non-negative integers n.
Note that, since there are $m_1 + m_2 + \cdots + m_t = k$ coefficients of these polynomials in total, these coefficients can be determined by solving the system of k equations resulting from the equations above for $n = 0, 1, \ldots, k-1$.

Problem 11.16. Let k be a positive integer, and let

$$a_n = \sum_{i=1}^{k} c_i a_{n-i}$$

be a homogeneous linear recurrence relation of depth k with initial conditions $a_0, a_1, \ldots, a_{k-1}$. Define the generating function

$$A(x) = \sum_{i=0}^{\infty} a_i x^i$$

and the polynomial

$$C(x) = \sum_{i=1}^{k} c_i x^i.$$

Prove that there exists a polynomial $B(x)$ of degree at most $k-1$, whose coefficients are determined by $c_1, c_2, \ldots, c_k$ and the initial conditions $a_0, a_1, \ldots, a_{k-1}$, such that

$$A(x)(1 - C(x)) = B(x).$$

Chapter 12

Group Theory

"It is difficult to give an idea of the vast extent of modern mathematics. The word "extent" is not the right one: I mean extent crowded with beautiful detail - not an extent of mere uniformity such as an objectless plain, but of a tract of beautiful country seen at first in the distance, but which will bear to be rambled through and studied in every detail of hillside and valley, stream, rock, wood, and flower."

– *Arthur Cayley*

"Mathematics is rich and beautiful because birds give it broad visions and frogs give it intricate details... It is stupid to claim that birds are better than frogs because they see farther, or that frogs are better than birds because they see deeper. The world of mathematics is both broad and deep."

– *Freeman Dyson, Birds and Frogs*

"Mathematicians are something categorized as either "problem solvers" or "theorists."... a theory that does not lead to the solution of concrete and interesting problems is not worth having. Conversely, any really deep problem tends to stimulate the development of theory for its solution."

– *Michael Atiyah*

As a capstone to our study of combinatorics, we will look at a structure that encapsulates much about symmetry: groups. We will cross the bridge from elementary mathematics to higher mathematics by studying a bit of group theory in abstract algebra. After building up some machinery about groups, we will prove Burnside's lemma. Using Burnside's lemma, we will solve problems about counting necklaces built out of beads such that rotations of the same necklace are considered to be non-distinct.

12.1 Groups and Burnside

Definition 12.1. As briefly described in Volume 1, a **group** is an ordered pair $(G, *)$ of a set G and a binary operation $* : G \times G \to G$ (typically written in infix notation or without any symbol at all for the operation) such that the following properties hold:

1. Associativity: For all $a, b, c \in G$,
$$(a * b) * c = a * (b * c).$$

2. Identity: There exists an $e \in G$ such that, for all $a \in G$,
$$e * a = a * e = a.$$
We showed in Volume 1 that such an identity element must be unique.

3. Inverses: For each $a \in G$, there exists an $a^{-1} \in G$ such that
$$a * a^{-1} = a^{-1} * a = e.$$
We showed in Volume 1 that such an inverse element of a is unique, due to associativity.

Note that commutativity is not a requirement on the group operation, though, when it holds, the group is called **abelian**. Our focus will be on finite G.

Definition 12.2. Given a set X and a group G, a **group action** of G on X is a function

$$\begin{aligned}\phi : G \times X &\to X \\ (g, x) &\mapsto \phi(g, x) = gx\end{aligned}$$

such that the following properties hold:

1. Identity: for all $x \in X$, $ex = x$, where e is the identity of G.
2. Compatibility: for all $x \in X$ and for all $g, h \in G$,
$$(gh)x = g(hx).$$

Definition 12.3. Given a group G, a set X, and an action of G on X, we define and denote:

1. For each $x \in X$, the **orbit** of x is
$$\text{Orb}(x) = \{gx : g \in G\}.$$
The set of orbits of elements of X is
$$X/G = \{\text{Orb}(x) : x \in X\}.$$
2. For each $x \in X$, the **stabilizer** of x is
$$\text{Stab}(x) = \{g \in G : gx = x\}.$$
3. For each $g \in G$, the **fix** of g is
$$\text{Fix}(g) = \{x \in X : gx = x\}.$$

Theorem 12.4. The set of orbits X/G of a set X under the action of a group G forms a partition of X, meaning the set of orbits consists of disjoint sets whose union is X.

Proof. Although we could show a proof that takes advantage of the fact that an equivalence relation induces a partition, we will perform a detailed manual proof in order to stay close to the fundamental definitions. Firstly, it is clear that $\bigcup\limits_{x \in X} \text{Orb}(x) = X$ because $x = ex \in \text{Orb}(x)$ shows that

$$X \subseteq \bigcup_{x \in X} \text{Orb}(x),$$

and the reverse inclusion

$$\bigcup_{x \in X} \text{Orb}(x) \subseteq X$$

is true by virtue of the fact that every orbit is a subset of X.
Secondly, we will prove that every pair of orbits $\text{Orb}(x)$ and $\text{Orb}(y)$ is either disjoint or identical by proving that, if there is a shared element $z \in \text{Orb}(x) \cap \text{Orb}(y)$, then $\text{Orb}(x) = \text{Orb}(y)$. Supposing $z \in \text{Orb}(x)$ and $z \in \text{Orb}(y)$, there exist $g, h \in G$ such that

$$gx = z = hy.$$

Then

$$x = g^{-1}gx = g^{-1}hy \in \text{Orb}(y).$$

So there exists a $j \in \text{Orb}(y)$ such that $x = jy$. Then, for every $g \in G$,

$$gx = gjy \in \text{Orb}(y),$$

and so $\text{Orb}(x) \subseteq \text{Orb}(y)$. By a symmetrical argument, $\text{Orb}(y) \subseteq \text{Orb}(x)$. Antisymmetry of sets yields

$$\text{Orb}(x) = \text{Orb}(y).$$

□

Theorem 12.5. If S is a set and $\sim$ is an equivalence relation on S, then the resulting set of equivalence classes forms of partition of S, meaning the union of the equivalence classes equals S and the classes are pairwise disjoint.

Proof. For the union property, every element $x \in S$ is in the equivalence class of itself

$$[x] = \{y \in S : y \sim x\}$$

due to the reflexive property $x \sim x$. Then

$$S \subseteq \bigcup_{x\in S} [x],$$

and the reverse inclusion

$$\bigcup_{x\in S} [x] \subseteq S$$

is true because every equivalence relation $[x]$ is a subset of S.
For the disjointedness property, suppose $z \in [x] \cap [y]$, so z is a shared element of two equivalence classes. We will show that the two classes collapse into each other, as in they are equal. By transitivity, $x \sim z$ and $z \sim y$ lead to $x \sim y$. Then $[x] \subseteq [y]$ and $[y] \subseteq [x]$, so $[x] = [y]$ by antisymmetry. □

Theorem 12.6 (Orbit-Stabilizer theorem)**.** If G is a finite group and X is a set on which G acts, then, for each $x \in X$,

$$|\text{Orb}(x)| \cdot |\text{Stab}(x)| = |G|.$$

Proof. We define a relation on G by

$$g \sim h \iff gx = hx.$$

It is immediately verifiable that reflexivity, symmetry, and transitivity hold, so it is an equivalence relation. Let the distinct equivalence classes be $E_1, E_2, \ldots, E_m$, which we know to form a partition of G by Theorem 12.5. Note that, since $\text{Orb}(x) = \{gx : g \in G\}$ is the definition of the orbit of x, and the equivalence relation is defined as

$$g \sim h \iff gx = hx,$$

the number of equivalence classes is the number of distinct elements of $\text{Orb}(x)$, so

$$m = |\text{Orb}(x)|.$$

Due to the fact that the E_i form a partition of G, we find that

$$|G| = \sum_{i=1}^{m} |E_i|.$$

We claim that, for each $i \in [m]$,

$$|E_i| = |\text{Stab}(x)|.$$

To this end, let $i \in [m]$ and $g \in E_i$. Then, for all $h \in G$,

$$h \in E_i \iff gx = hx \iff x = g^{-1}hx \iff g^{-1}h \in \text{Stab(x)}.$$

We define

$$g \cdot \text{Stab}(x) = \{gj : j \in \text{Stab}(x)\}.$$

Then

$$\begin{aligned} g^{-1}h \in \text{Stab}(x) &\iff \exists j \in \text{Stab}(x) : g^{-1}h = j \\ &\iff \exists j \in \text{Stab}(x) : h = gj \\ &\iff h \in g \cdot \text{Stab}(x). \end{aligned}$$

So,

$$h \in E_i \iff h \in g \cdot \mathrm{Stab}(x),$$

leading to

$$|E_i| = |g \cdot \mathrm{Stab(x)}|.$$

But we can produce a bijection from Stab(x) to $g \cdot$ Stab(x) easily using multiplication by g, so the bijection principle says that

$$|E_i| = |g \cdot \mathrm{Stab(x)}| = |\mathrm{Stab}(x)|.$$

Therefore,

$$|G| = \sum_{i=1}^{m} |E_i| = \sum_{i=1}^{m} |\mathrm{Stab}(x)| = m \cdot |\mathrm{Stab}(x)| = |\mathrm{Orb}(x)| \cdot |\mathrm{Stab}(x)|.$$

□

Theorem 12.7 (Burnside's lemma). Let G be a finite group acting on a finite set X. Then the number of orbits of X is given by

$$|X/G| = \frac{1}{|G|} \cdot \sum_{g \in G} |\mathrm{Fix}(g)|.$$

Proof. By the discrete Fubini's principle,

$$\begin{aligned}\sum_{g \in G} |\mathrm{Fix}(g)| &= \sum_{g \in G} \{x \in X : gx = x\} \\ &= |\{(g, x) \in G \times X : gx = x\}| \\ &= \sum_{x \in X} |\{g \in G : gx = x\}| \\ &= \sum_{x \in X} |\mathrm{Stab}(x)|.\end{aligned}$$

By the orbit-stabilizer theorem (Theorem 12.6),

$$\sum_{g \in G} |\mathrm{Fix}(g)| = \sum_{x \in X} |\mathrm{Stab}(x)| = \sum_{x \in X} \frac{|G|}{|\mathrm{Orb}(x)|} = |G| \cdot \sum_{x \in X} \frac{1}{|\mathrm{Orb}(x)|}.$$

By the fact that the set of orbits X/G of X partitions X (Theorem 12.4), the sum on the right side is

$$\sum_{x \in X} \frac{1}{|\mathrm{Orb}(x)|} = \sum_{E \in X/G} \sum_{x \in E} \frac{1}{|\mathrm{Orb}(x)|} = \sum_{E \in X/G} \sum_{x \in E} \frac{1}{|E|} = \sum_{E \in X/G} |E| \cdot \frac{1}{|E|} = \sum_{E \in X/G} 1 = |X/G|.$$

Therefore,

$$\sum_{g \in G} |\mathrm{Fix}(g)| = |G| \cdot |X/G|.$$

□

Those who are interested in a far-reaching generalization of Burnside's lemma are encouraged to study Pólya's enumeration theorem.

12.2 Counting Necklaces

In Theorem 3.23, we counted circular permutations, where each element in the permutation is distinct. Now we will consider the problem of counting necklaces where there is an unlimited number of each bead colour available.

Definition 12.8. Given a permutation $\pi : [n] \to [n]$ (i.e. a bijection), we can denote the function in **cycle decomposition notation**. For example, the permutation

$$\begin{bmatrix} 1 & 2 & 3 & 4 & 5 & 6 \\ 3 & 5 & 4 & 1 & 6 & 2 \end{bmatrix}$$

may be denoted by $(1\ 3\ 4)(2\ 5\ 6)$ because

$$1 \mapsto 3 \mapsto 4 \mapsto 1,$$
$$2 \mapsto 5 \mapsto 6 \mapsto 2.$$

Definition 12.9. Given positive integers n and k, an (n, k)**-necklace** is a circular arrangement of n beads, each of which are chosen from an unlimited supply of each of k colours. The key point is that rotations of the same necklace are considered to be the same, but reflections are distinct.

Example 12.10. Given that k is a positive integer, find a closed expression for the number of $(6, k)$-necklaces.

Solution. Let X be the set of the k^6 assignments of colours to the 6 beads. Let G be the group of rotations acting on X. If $\sigma \in G$ is the "generator" rotation that causes a counterclockwise rotation by $\frac{2\pi}{6}$ radians, then the elements of G consist of applying σ up to and including 6 times:

$$G = \{\sigma, \sigma^2, \sigma^3, \sigma^4, \sigma^5, \sigma^6 = e\},$$

where σ^i applies the σ rotation i times. Each distinct necklace is an orbit of X, so we are seeking to count the number of orbits $|X/G|$, which we will find using Burnside's lemma (Theorem 12.7). To do that, we will need to compute $|\text{Fix}(\sigma^i)|$ for each $i \in [6]$, to which end we will study the cycle decomposition notation of each σ^i:

$$\begin{aligned} \sigma &= (1\ 2\ 3\ 4\ 5\ 6), \\ \sigma^2 &= (1\ 3\ 5)(2\ 4\ 6), \\ \sigma^3 &= (1\ 4)(2\ 5)(3\ 6), \\ \sigma^4 &= (1\ 5\ 3)(2\ 6\ 4), \\ \sigma^5 &= (1\ 6\ 5\ 4\ 3\ 2), \\ e = \sigma^6 &= (1)(2)(3)(4)(5)(6) \end{aligned}$$

Since

$$\text{Fix}(\sigma^i) = \{x \in X : \sigma^i x = x\},$$

computing $|\text{Fix}(\sigma^i)|$ means counting the number of ways that the colours in each cycle of the cycle decomposition of σ^i are the same. If there are $c(\sigma^i)$ cycles in the cycle decomposition notation of σ^i and there are k colours available, then

$$|\text{Fix}(\sigma^i)| = k^{c(\sigma^i)}.$$

By Burnside's lemma,

$$\begin{aligned} |X/G| &= \frac{1}{|G|} \cdot \sum_{i=1}^{6} |\text{Fix}(\sigma^i)| \\ &= \frac{1}{6}(k + k^2 + k^3 + k^2 + k + k^6) \\ &= \frac{k(k+1)(k^2+k+1)(k^2-2k+2)}{6}. \end{aligned}$$

□

Theorem 12.11. If n and k are positive integers, then the number of (n, k)-necklaces is given by

$$\frac{1}{n} \cdot \sum_{i=1}^{n} k^{\gcd(n,i)}.$$

Proof. Following the logic of Example 12.10, it suffices to prove that, if σ is the rotation by $\frac{2\pi}{n}$ radians, then the number of cycles in the cycle decomposition notation of σ^i is $\gcd(n, i)$. In fact, notice that the cycles in each cycle decomposition seem to have the same length in the example. So it suffices to show that the length of each cycle in the decomposition of σ^i is $\frac{n}{\gcd(n,i)}$, because then the number of cycles would be

$$\frac{n}{\left(\frac{n}{\gcd(n,i)}\right)} = \gcd(n, i),$$

as desired.
Changing the notation a bit to take advantage of modular arithmetic, we will show that if

$$\sigma = (0\ 1\ 2\ \ldots\ n-1)$$

is an n-cycle of the set $\{0, 1, 2, \cdots, n-1\}$, then σ^i has a cycle decomposition notation with $\frac{n}{\gcd(n,i)}$ cycles, each with length $\gcd(n, i)$. Note that σ^i sends $x \mapsto x + i \pmod{n}$ for each $i = 0, 1, 2, \ldots, n-1$, where the output is reduced to its least non-negative representative modulo n. By the definition of a cycle, if t is the length of the decomposition's cycle containing x, then t is the least positive integer k satisfying

$$x + ki \equiv x \pmod{n}.$$

Equivalently, $ki \equiv 0 \pmod{n}$ or $n \mid ki$ which is a condition that is independent of x. Therefore,

$$t = \min\{k \in \mathbb{Z}_+ : n \mid ki\} = \min\left\{k \in \mathbb{Z}_+ : \frac{n}{\gcd(n,i)} \,\middle|\, k \cdot \frac{i}{\gcd(n,i)}\right\}.$$

Since

$$\gcd\left(\frac{n}{\gcd(n,i)}, \frac{i}{\gcd(n,i)}\right) = 1,$$

Gauss's divisibility lemma from Volume 3 makes the condition equivalent to $\frac{n}{\gcd(n,i)} \,\Big|\, k$. Therefore,

$$t = \min\left\{k \in \mathbb{Z}_+ : \frac{n}{\gcd(n,i)} \,\middle|\, k\right\} = \frac{n}{\gcd(n,i)},$$

since the minimal positive integer divisible by some integer is that integer itself. □

Problem 12.12. If p is a prime and k is a positive integer, then find a closed formula for the number of (p, k)-necklaces. For those who are familiar with modular arithmetic, do you see a link with Fermat's little theorem from Volume 3?

Problem 12.13. Determine the number of distinct directed graphs on 3 vertices such that two edges between two vertices are allowed if they are in the two different directions, and there are no loops. Graphs are considered to be the same if they are isomorphic, meaning relabelling the vertices does not alter the underlying edge structure.

Now we will consider what happens when there are two colours, each with a specified number of beads. It will be helpful to have familiarity with some of the later material on number theory.

Lemma 12.14. Let b, c be positive integers such that $c \mid b$. Then, the number of integers $j \in [b]$ such that $(b, j) = c$ is given by $\varphi\left(\frac{b}{c}\right)$. Here, (b, c) refers to $\gcd(b, j)$, which is a convention that we adopted for simplicity of notation.

Proof. Since $c = (b, j)$ and $(b, j) \mid j$ by definition, we get $c \mid j$. Then, $\frac{b}{c}$ and $\frac{j}{c}$ are positive integers, so we can make the well-defined statement

$$(b, j) = c \iff \left(\frac{b}{c}, \frac{j}{c}\right) = 1.$$

So, it suffices to count $j \in [b]$ such that $\frac{b}{c}$ and $\frac{j}{c}$ are coprime. Note that such $j \in [b]$ are in bijection with $\frac{j}{c} \in \left[\frac{b}{c}\right]$. By the definition of Euler's totient function, the number of $\frac{j}{c}$ in $\left[\frac{b}{c}\right]$ that are coprime to $\frac{b}{c}$ is $\varphi\left(\frac{b}{c}\right)$. $\square$

Problem 12.15. For each pair of positive integers n and n, prove that

$$\sum_{i=1}^{n} k^{\gcd(n,i)} = \sum_{d \mid n} \varphi(d) k^{\frac{n}{d}},$$

where the sum on the right ranges over all positive divisors d of n. This gives an alternative expression for the solution to the necklace-counting question (Theorem 12.11) by collecting similar powers.

Theorem 12.16. The number of binary necklaces, as in necklaces with beads that are each black or white, where there are $n \geq 1$ total beads and exactly m black beads for some integer $1 \leq m \leq n$, is given by

$$\frac{1}{n} \sum_{d \mid (n,m)} \varphi(d) \binom{n/d}{m/d}.$$

Proof. By Burnside's lemma, if G is the group of rotations on n beads, and Y is the set of all n-beaded necklaces with exactly k black beads and $n - k$ white beads, then

$$|Y/G| = \frac{1}{|G|} \sum_{g \in G} |\text{Fix}(g)| = \frac{1}{n} \sum_{i=1}^{n} |\text{Fix}(\sigma^i)|.$$

So, our current goal is to compute

$$|\text{Fix}(\sigma^i)| = |\{y \in Y : \sigma^i y = y\}|.$$

Recall from Example 12.10 and Theorem 12.11 that σ^i has (n, i) cycles, each of which has $\frac{n}{(n,i)}$ beads, and the colour of all beads in each individual cycle is constant (either black or white). Since there are m black beads, it is necessary that $\frac{n}{(n,i)} \,\Big|\, m$; in this case, the number of cycles that are black is

$$\frac{m}{\left[\frac{n}{(m,i)}\right]} = \frac{m(n,i)}{n}.$$

Since these black cycles can be chosen arbitrarily from any of the total (n, i) cycles, we get

$$|\text{Fix}(\sigma^i)| = \binom{(n,i)}{\frac{m(n,i)}{n}}.$$

Now, it is a matter of simplifying the sum of such terms, as i ranges from 1 to n. We need a better grasp on the criterion $\left.\frac{n}{(n,i)}\,\right|\, m$, since i is difficult to access when it is tucked away inside a gcd. To that end, notice that

$$\left.\frac{n}{(n,i)}\,\right|\, m \iff n \mid m(n,i) \iff n \mid (nm, im) \iff n \mid nm, im \iff n \mid im.$$

We use this and Gauss's divisibility lemma from Volume 3 to get

$$\left.\frac{n}{(n,i)}\,\right|\, m \iff n \mid im \iff \left.\frac{n}{(n,m)}\,\right|\, i \cdot \frac{m}{(n,m)} \iff \left.\frac{n}{(n,m)}\,\right|\, i.$$

Then,

$$\sum_{i=1}^{n} |\mathrm{Fix}(\sigma^i)| = \sum_{\substack{i\in[n] \\ \frac{n}{(n,i)}|m}} \binom{(n,i)}{\frac{m(n,i)}{n}} = \sum_{\substack{i\in[n] \\ \frac{n}{(n,m)}|i}} \binom{(n,i)}{\frac{m(n,i)}{n}}.$$

The conjunction of $i \in [n]$ and $\left.\frac{n}{(n,m)}\,\right|\, i$ means we are iterating over $i \in [n]$ that are multiples of $\frac{n}{(n,m)}$. There are $\frac{n}{\left[\frac{n}{(n,m)}\right]} = (n,m)$ such integers, specifically $i = \frac{nj}{(n,m)}$ for $j = 1, 2, \ldots, (n,m)$. Letting $b = (n,m)$, we get

$$\sum_{i=1}^{n} |\mathrm{Fix}(\sigma^i)| = \sum_{\substack{i\in[n] \\ \frac{n}{b}|i}} \binom{(n,i)}{\frac{m(n,i)}{n}} = \sum_{j=1}^{b} \binom{\left(n, \frac{nj}{b}\right)}{\frac{m\left(n, \frac{nj}{b}\right)}{n}} = \sum_{j=1}^{b} \binom{\frac{n}{b}(b,j)}{\frac{m}{b}(b,j)}.$$

As j iterates over the integers from 1 to b, the distinct values taken on by (b,j) are all the divisors c of $b = (n,m)$. Using Lemma 12.14, we know that, for each divisor $c = (b,j)$ of b, the number of times that

$$\binom{\frac{n}{b}(b,j)}{\frac{m}{b}(b,j)} = \binom{\frac{nc}{b}}{\frac{mc}{b}}$$

appears in the sum as j goes from 1 to b is $\varphi\left(\frac{b}{c}\right)$. Therefore,

$$|Y/G| = \frac{1}{n}\sum_{c|b} \varphi\left(\frac{b}{c}\right)\binom{\frac{nc}{b}}{\frac{mc}{b}} = \frac{1}{n}\sum_{cd=b} \varphi\left(\frac{b}{c}\right)\binom{n/(b/c)}{m/(b/c)} = \frac{1}{n}\sum_{d|(n,m)} \varphi(d)\binom{n/d}{m/d},$$

as desired.

As a side note, the count is more difficult and based on parity casework if reflections are also considered to be non-distinct. We have not included this variant. □

Appendices

Appendix A

Matrices

"When I look at the history of mathematics, I see a succession of illogical jumps, improbable coincidences, jokes of nature."

– *Freeman Dyson, Birds and Frogs*

"Mathematics is an experimental science, and definitions do not come first, but later on. They make themselves, when the nature of the subject has developed itself."

– *Oliver Heaviside, On operations in physical mathematics II*

We will outline here the definitions and basic properties of matrices that are used in the book.

Definition A.1. If m and n are positive integers, then, informally, an $m \times n$ **matrix** is an array or table of numbers with m rows and n columns. Formally, it is an indexing function whose indexing set is $[m] \times [n]$.

Definition A.2. The **matrix multiplication** of 2×2 matrices is defined by

$$\begin{pmatrix} a & b \\ c & d \end{pmatrix} \cdot \begin{pmatrix} p & q \\ r & s \end{pmatrix} = \begin{pmatrix} ap + br & aq + bs \\ cp + dr & cq + ds \end{pmatrix}.$$

Similarly, a 2×2 matrix times a column vector is computed as

$$\begin{pmatrix} a & b \\ c & d \end{pmatrix} \cdot \begin{pmatrix} p \\ q \end{pmatrix} = \begin{pmatrix} ap + bq \\ cp + dq \end{pmatrix}.$$

Matrix multiplication is associative, but not commutative. Positive integer **powers** of a 2×2 matrix are defined as repeated multiplication, just like the exponentiation of integers. The multiplication of larger matrices follows a similar definition. The multiplication of a matrix by a scalar (for us, a scalar is a real number but a more general definition involving field elements exists in abstract algebra), called **scalar multiplication**, is defined as

$$\alpha \cdot \begin{pmatrix} a & b \\ c & d \end{pmatrix} = \begin{pmatrix} \alpha a & \alpha b \\ \alpha c & \alpha d \end{pmatrix}.$$

We need matrix multiplication in the proof of the barycentric shoelace formula in Volume 4. The multiplication of a matrix with a column vector is useful as well for succinctly expressing a system of equations in the proof of the uniqueness of bivariate quadratic representations of conics in Volume 4. The powers of a matrix are used in a proof of Cassini's identity for the Fibonacci numbers in Volume 2.

Definition A.3. We call

$$I = \begin{pmatrix} 1 & 0 \\ 0 & 1 \end{pmatrix}$$

the 2×2 **identity matrix** because $IM = MI$ for every 2×2 matrix M, as the reader should verify. The identity matrix is like the number 1 in the multiplication of real numbers, since multiplying any matrix by I gives back that matrix.

Definition A.4. We will not define the **determinant** of an $n \times n$ matrix in general but the following formulas show how to compute the determinants of 2×2 and 3×3 matrices:

- The determinant of a 2×2 matrix is the real number

$$\det \begin{pmatrix} a & b \\ c & d \end{pmatrix} = ad - bc.$$

- The determinant of a 3×3 matrix is the real number

$$\det \begin{pmatrix} a & b & c \\ d & e & f \\ g & h & i \end{pmatrix} = aei + bfg + cdh - ceg - bdi - afh.$$

Theorem A.5. One way of remembering how to compute the 3×3 determinant, called the **Rule of Sarrus**, is shown by the following diagram:

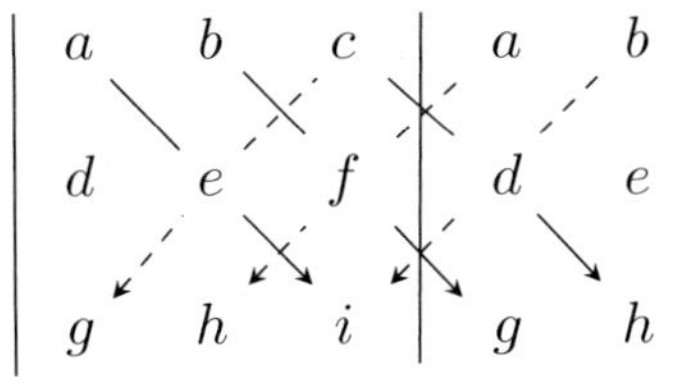

Theorem A.6. For 2×2 matrices M and N,

$$\det(M \cdot N) = \det(M) \cdot \det(N).$$

This is called the multiplicative property of the determinant.

Proof. We will use the matrices $M = \begin{pmatrix} a & b \\ c & d \end{pmatrix}$ and $N = \begin{pmatrix} p & q \\ r & s \end{pmatrix}$. Their product is

$$M \cdot N = \begin{pmatrix} ap + br & aq + bs \\ cp + dr & cq + ds \end{pmatrix}.$$

We find that

$$\begin{aligned} \det(M \cdot N) &= (ap + br)(cq + ds) - (aq + bs)(cp + dr) \\ &= adps + bcqr - adqr - bcps \\ &= (ad - bc)(ps - qr) \\ &= \det(M) \cdot \det(N). \end{aligned}$$

□

We will need the following properties that hold for the determinants of all square matrices, though we will need to apply them only to 2×2 and 3×3 matrices.

Theorem A.7. The facts below about determinants may be verified using the algebraic definitions (Definition A.4), albeit such direct proofs are tedious:

1. A scalar can be factored out of a single row or a single column of the determinant to outside the determinant without altering the value of the determinant. Equivalently, multiplying every element of one row or every element of one column by the same number results in multiplying the determinant by the same number.

2. Adding or subtracting (a scalar multiple) of a row from another row preserves the determinant. The same is true for the corresponding statement about columns.

3. Swapping two rows with each other or two columns with each other changes the sign of the determinant.

4. The determinant is preserved under taking the transpose of the matrix, meaning the element in each row i and column j is exchanged with the element in row j and column i. This operation is equivalent to reflecting the matrix across the diagonal that runs from the top-left to the bottom-right.

The determinant appears in a proof of the extended Euclidean algorithm in Volume 3, when studying lines from the perspective of coordinates in Volume 4, and in the derivation of various shoelace-style area formulas in Volume 4. Also, the Vandermonde matrix from Volume 2 and its determinant are relevant to the study of polynomials in Volume 1, and again in the theory of linear recurrences in Volume 2.

Appendix B

Solutions

"[Gauss] is like the fox, who effaces his tracks in the sand with his tail."

– *Niels Henrik Abel*

Solution 1.9. Let X be finite and $Y \subsetneq X$. Since X is a non-empty finite set, there exists a positive integer n such that $X \approx [n]$. Since Y is a proper subset of X, Lemma 1.5 implies that either $Y = \varnothing$ or $Y \approx [m]$ for some positive integer $m < n$. In the former case $|Y| = 0 < n = |X|$, and in the latter case $|Y| = m < n = |X|$. Either way, $|Y| < |X|$.

Solution 1.11. The second part will follow from the first, so we will prove the assertions in sequence:

1. If $X = \varnothing$ then we are asked to prove that $|\{x_0\}| = 1$, which is true because $\{x_0\} \approx [1]$. Now suppose X is a non-empty finite set. Then there exists a bijection $f : X \to [n]$. We simply extend this bijection by defining $g : X \cup \{x_0\} \to [n+1]$ as

$$g(x) = \begin{cases} f(x) & \text{if } x \in X \\ n+1 & \text{if } x = x_0 \end{cases},$$

which is also bijection. Thus, $X \cup \{x_0\}$ is finite and $|X \cup \{x_0\}| = n + 1 = |X| + 1$.

2. For the second assertion, let $Z = Y \backslash \{y_0\}$. Then Z is a proper subset of Y, which makes Z finite. Moreover, y_0 is not an element of Z. By the previous part of the problem, this means $Z \cup \{y_0\}$ is finite and $|Z \cup \{y_0\}| = |Z| + 1$. By definition, $Z \cup \{y_0\} = Y$ and $Z = Y \backslash \{y_0\}$, so the equation that we derived is equivalent to $|Y| = |Y \backslash \{y_0\}| + 1$, which is further equivalent to $|Y \backslash \{y_0\}| = |Y| - 1$.

Solution 1.13. We start with 0 and then toggle between positive and negative integers:

$$0, 1, -1, 2, -2, 3, -3, 4, -4 \ldots,$$

which produces a bijection from $\mathbb{Z}_+$ to $\mathbb{Z}$. The interested reader can construct an explicit piecewise bijection.

Solution 1.16. Let X be countably infinite and suppose for contradiction that X is finite. Then $X \approx \mathbb{Z}_+$ and $X \approx [n]$ for some positive integer n. As a result, $\mathbb{Z}_+ \approx [n]$, meaning $\mathbb{Z}_+$ is finite. We will show that this conclusion is impossible by using Lemma 1.6. Let $Y = \mathbb{Z}_+ \backslash \{1\}$. Since Y is a non-empty proper subset of $\mathbb{Z}_+$, and since we have arrived at the hypothetical conclusion that $\mathbb{Z}_+$ is a non-empty finite set, Lemma 1.6 says that $\mathbb{Z}_+ \not\approx Y$. However, the function $f : \mathbb{Z}_+ \to Y$, defined by

$$f : n \mapsto n + 1$$

is a bijection, which proves that $\mathbb{Z}_+ \approx Y$. This is a contradiction.

Solution 1.19. Since a determines b and b determines c, it all boils down to the choice of a. At first, it might seem that the possibilities for a are all the given digits. However, a closer look reveals that a cannot be 8 or 9 because it would prevent at least one of $b = a + 1$ or $c = a + 2$ from being a digit from the given set. Thus, the possibilities for a are $0, 1, 2, 3, 4, 5, 6, 7$, and so the set of possible lists abc is

$$\{012, 123, 234, 345, 456, 567, 678, 789\}.$$

Solution 1.20. An important idea that occurs time and again in combinatorics is that choosing three distinct real numbers automatically leads to exactly one way of placing them in increasing order (or decreasing order). So we just need to find all 3-element subsets of $\{1, 2, 3, 4, 5\}$ and write each subset as an increasing list.
It is not easy to work with 3 symbols, as we might not be sure at the end of the process that we have found all possibilities. However, choosing 3 different symbols to be in a subset is the same as excluding 2 different symbols from this subset. So we begin by finding the set of all possible choices of 2 different symbols:

$$\{12, 13, 14, 15, 23, 24, 25, 34, 35, 45\}.$$

This leads to the set of all possible choices of 3 different symbols:

$$\{345, 245, 235, 234, 145, 135, 134, 125, 124, 123\}.$$

For example, we found the first element 345 by excluding 12 from 12345.

Solution 1.22. Since order does not matter, we can use the number of 0's in the set as the deciding factor. This is because fixing the number of 0's yields the number of 1's. The number of 0's is $0, 1, 2, 3$ or 4. In these respective cases, the number of 1's is $4, 3, 2, 1$ or 0. Thus, the set of possible multisets is

$$\{1111, 1110, 1100, 1000, 0000\},$$

where we have informally written each 4-multiset $\langle a, b, c, d\rangle$ as $abcd$.

Solution 1.35. By the addition principle, A, B, C are all finite and

$$|X| = |A| + |B| + |C|.$$

By the bijection principle, $B \approx C$ implies that $|B| = |C|$. Thus, $|X| = |A| + 2|B|$, which is equivalent to the desired equation.
For the application to dice, we know that the probability of the blue die's number being strictly higher than the green die's number is exactly the same as the probability of the green die's number being strictly higher than the blue die's number. There is also the "in between" third case of the two numbers being equal, which can occur in 6 ways, out of a total $6 \cdot 6 = 36$ possible pairs of rolls. Therefore, the answer is

$$\frac{1 - \frac{6}{36}}{2} = \frac{1 - \frac{1}{6}}{2} = \frac{\left(\frac{5}{6}\right)}{2} = \frac{5}{12}.$$

Solution 1.39. The result makes sense intuitively because even the slightest overlap between a pair of the sets would result in the full union having a sub-optimal cardinality. We will do a proof by induction on $n \geq 1$. In the base case $n = 1$, the result is obviously true. So we assume that the result holds for some integer $n \geq 1$. Let $\langle A_1, A_2, \ldots, A_n, A_{n+1}\rangle$ be a multiset of $n + 1$ finite sets. Let

$$B = A_1 \cup A_2 \cup \cdots \cup A_n.$$

By the principle of inclusion-exclusion for two sets,

$$|B \cup A_{n+1}| = |B| + |A_{n+1}| - |B \cap A_{n+1}| \leq |B| + |A_{n+1}|,$$

with equality holding if and only if $|B \cap A_{n+1}| = 0$, which is true if and only if $B \cap A_{n+1} = \varnothing$. By the definition of B and the distributive property of set intersection over set union,

$$\begin{aligned}\varnothing &= B \cap A_{n+1}\\ &= (A_1 \cup A_2 \cup \cdots \cup A_n) \cap A_{n+1}\\ &= (A_1 \cap A_{n+1}) \cup (A_2 \cap A_{n+1}) \cup \cdots \cup (A_n \cap A_{n+1}).\end{aligned}$$

For each index $1 \le i \le n$, the intersection $(A_i \cap A_{n+1})$ is a subset of this union, so each $A_i \cap A_{n+1}$ is also empty. Conversely, if each $A_i \cap A_{n+1}$ is empty, then their union $B \cap A_{n+1}$ is empty. So equality holds in the above inequality if and only if each $A_i \cap A_{n+1}$ is empty.
All we have to do now is invoke the induction hypothesis. Since $\langle A_1, A_2, \ldots, A_n \rangle$ is a multiset of n finite sets, the induction hypothesis states that

$$|B| \le |A_1| + |A_2| + \cdots + |A_n|,$$

with equality holding if and only if $\langle A_1, A_2, \ldots, A_n \rangle$ is pairwise disjoint. Thus,

$$\begin{aligned}|A_1 \cup A_2 \cup \cdots \cup A_n \cup A_{n+1}| &= |B \cup A_{n+1}| \\ &= |B| + |A_{n+1}| - |B \cap A_{n+1}| \\ &\le |B| + |A_{n+1}| \\ &\le |A_1| + |A_2| + \cdots + |A_n| + |A_{n+1}|.\end{aligned}$$

Equality holds if and only if equality holds at each step. Combining the conditions that $A_i \cap A_{n+1} = \varnothing$ for each index $1 \le i \le n$ and that $\langle A_1, A_2, \ldots, A_n \rangle$ is pairwise disjoint, the equality condition for the case of $n+1$ sets is that $\langle A_1, A_2, \ldots, A_n, A_{n+1} \rangle$ is pairwise disjoint. This completes the induction.

Solution 2.4. Since X and Y are non-empty finite sets, there exist positive integers n and m such that $X \approx [n]$ and $Y \approx [m]$. We may assume without loss of generality that $n < m$ because a symmetrical argument exists for the case where $n > m$. From $X \approx [n]$ we obtain a bijection $g : X \to [n]$, and from $[m] \approx Y$ we obtain a bijection $h : [m] \to Y$. Moreover, the identity function $\mathrm{Id}_{[n]} : [n] \to [m]$ is an injection. Thus, the composition

$$(h \circ \mathrm{Id}_{[n]} \circ g) : X \to Y$$

is an injection because every bijection is an injection and a composition of injections is an injection. As a consequence of the injection-surjection lemma (Lemma 2.3), there exists a surjection $k : Y \to X$.

Solution 2.7. Let X and Y be non-empty finite sets such that $|X| = |Y|$.

1. Let $f : X \to Y$ be an injection. For the sake of contradiction, suppose f is not a surjection. Then there exists an element $y \in Y$ that is not in the range of f. The idea is to extend f to have y in its range by letting z be an element outside X that gets mapped to y. That is, we define the new domain $Z = X \sqcup \{z\}$ and then $g : Z \to Y$ by

$$g(x) = \begin{cases} f(x) & \text{if } x \in X \\ y & \text{if } x = z \end{cases},$$

which is easily seen to inherit the injectivity of f. Then

$$|Z| = |X \sqcup \{z\}| = |X| + 1 = |Y| + 1 > |Y|,$$

due to which the pigeonhole principle implies that g is not injective. This is a contradiction.

2. Let $f : X \to Y$ be a surjection. For the sake of contradiction, suppose f is not injective. Then, there exist two elements $x_1, x_2 \in X$ such that $x_1 \neq x_2$ and $f(x_1) = f(x_2)$. If X is a singleton, then we have already reached a contradiction. If not, we keep moving forward. In contrast with how we extended the domain of f in the previous part, the idea here is to restrict f to the domain $Z = X \backslash \{x_2\}$ (which still contains x_1 and so Z is non-empty). That is, we define $g : Z \to Y$ as $g = f|_Z$, which is seen to inherit the surjectivity of f since f maps x_1 and x_2 to the same output. Then

$$|Z| = |X \backslash \{x_2\}| = |X| - 1 = |Y| - 1 < |Y|,$$

due to which the reverse pigeonhole principle implies that g is not surjective. Again, this is a contradiction.

Regardless of whether we assume $f : X \to Y$ is an injection or surjection, we have proven that it is both. Thus, f is bijective in either case.
For the corollary, suppose S is a non-empty finite set of cardinality n. Then the definition of a permutation implies that $f : [n] \to S$ is a permutation of S if and only if f is an injection. Since $|S| = n = |[n]|$, what we have proven above implies that if f is an injection then f is a surjection, proving that f is a bijection. Of course, in the other direction, if f is a bijection, then it is an injection.

Solution 2.9. In each of the two cases, there are two sub-cases in each statement: X is finite or Y is finite. We will repeatedly use the injection-surjection lemma to flip our surjections into injections, so that the finite Schröder-Bernstein theorem can be applied.

1. If X is finite, then the existence of the surjection $f : X \to Y$ implies that Y is finite as well. On the other hand, if Y is finite then the existence of the surjection $g : Y \to X$ implies that X is finite as well. Either way, both X and Y are finite. Since Y is finite, we can flip g to produce an injection $p : X \to Y$, and, since X is finite, we can flip f to produce an injection $q : Y \to X$.

2. We already have a injection this case, and only need to flip the one surjection into an injection. If Y is finite, then the existence of the injection $h : X \to Y$ implies that X is finite as well. In the other sub-case, X is given as finite. Either way, X is finite, which allows us to flip the surjection $f : X \to Y$ to produce an injection $r : Y \to X$.

In either case, we can produce two injections mapping in opposite directions with least one of X, Y being finite, so the rest follows from applying the finite Schröder-Bernstein theorem.

Solution 2.11. Define the function

$$\begin{aligned} f : \{0, 1, \ldots, \lfloor \sqrt{p} \rfloor\}^2 &\to \{0, 1, \ldots, p-1\} \\ (x, y) &\mapsto ax - y \pmod p, \end{aligned}$$

where the output is reduced to its least non-negative representative modulo p. There are $(\lfloor \sqrt{p} \rfloor + 1)^2$ inputs and p outputs. By a property of the floor function, $\sqrt{p} < \lfloor \sqrt{p} \rfloor + 1$, which implies $p < (\lfloor \sqrt{p} \rfloor + 1)^2$, so there are strictly more inputs than outputs. By the pigeonhole principle, there exist distinct elements (x_1, y_1) and (x_2, y_2) of the domain of f such that

$$\begin{aligned} ax_1 - y_1 &\equiv ax_2 - y_2 \pmod p \\ a(x_1 - x_2) &\equiv y_1 - y_2 \pmod p. \end{aligned}$$

As a result, we get the hypothetical equivalence

$$x_1 \equiv x_2 \pmod p \iff y_1 \equiv y_2 \pmod p,$$

in which case (x_1, y_1) and (x_2, y_2) would be the same element of the domain, so neither is true. As such, we can let

$$\begin{aligned} x_0 &= x_1 - x_2 \not\equiv 0 \pmod p, \\ y_0 &= y_1 - y_2 \not\equiv 0 \pmod p. \end{aligned}$$

We simply have to check that x_0, y_0 lie in the stated intervals now. Indeed,

$$\begin{aligned} 0 < x_1 < \sqrt{p}, 0 < x_2 < \sqrt{p} &\implies -\sqrt{p} < x_1 - x_2 < \sqrt{p}, \\ 0 < y_1 < \sqrt{p}, 0 < y_2 < \sqrt{p} &\implies -\sqrt{p} < y_1 - y_2 < \sqrt{p}, \end{aligned}$$

and combining this with $p \nmid x_0$ and $p \nmid y_0$ yields the strict inequalities $0 < |x_0| < \sqrt{p}$ and $0 < |y_0| < \sqrt{p}$.

Solution 2.13. Let X and Y be non-empty finite sets. Assuming the strong pigeonhole principle, we want to show that, if $f : X \to Y$ is an injection, then $|X| \leq |Y|$. Since f is an injection, it means for all

$y \in Y$, $|f^{-1}(y)| \leq 1$. By the strong pigeonhole principle, there exists a $y \in Y$ such that $\left\lceil \frac{|X|}{|Y|} \right\rceil \leq |f^{-1}(y)|$. Combining this with the fact that $x \leq \lceil x \rceil$ for all real x, we get

$$\frac{|X|}{|Y|} \leq \left\lceil \frac{|X|}{|Y|} \right\rceil \leq |f^{-1}(y)| \leq 1$$

or $|X| \leq |Y|$.

Solution 2.15. Let X and Y be non-empty finite sets. Assuming the strong reverse pigeonhole principle, we want to show that, if $f : X \to Y$ is a surjection, then $|X| \geq |Y|$. Since f is a surjection, it means for all $y \in Y$, $|f^{-1}(y)| \geq 1$. By the strong reverse pigeonhole principle, there exists a $y \in Y$ such that $\left\lfloor \frac{|X|}{|Y|} \right\rfloor \geq |f^{-1}(y)|$. Combining this with the fact that $x \geq \lfloor x \rfloor$ for all real x, we get

$$\frac{|X|}{|Y|} \geq \left\lfloor \frac{|X|}{|Y|} \right\rfloor \geq |f^{-1}(y)| \geq 1$$

or $|X| \geq |Y|$.

Solution 2.17. Let k be positive integer and $n_1, n_2, \ldots, n_k$ be k positive integers.

1. Suppose, for contradiction, that for each index $1 \leq i \leq k$, hole number i receives fewer than n_i pigeons. Then the total number of pigeons in the holes is less than or equal to

$$\sum_{i=1}^{k} (n_i - 1) = \left(\sum_{i=1}^{k} n_i \right) - k,$$

which is strictly less than $\left(\sum_{i=1}^{k} n_i \right) - k + 1$, the prescribed number of pigeons. This is a contradiction.

2. Suppose, for contradiction, that for each index $1 \leq i \leq k$, hole number i receives more than n_i pigeons. Then the total number of pigeons in the holes is greater than or equal to

$$\sum_{i=1}^{k} (n_i + 1) = \left(\sum_{i=1}^{k} n_i \right) + k,$$

which is strictly greater than $\left(\sum_{i=1}^{k} n_i \right) + k - 1$, the prescribed number of pigeons. This is a contradiction again.

Solution 3.3. We break the possibilities up into cases according to the number of elements in the subset: $0, 1, 2, 3, 4$. There is only one subset with nothing in it, which is the empty set $\varnothing$. For subsets with one element, otherwise known as singletons, we simply choose one element which yields the four possibilities $\{a\}, \{b\}, \{c\}, \{d\}$. Skipping over subsets with two elements for a moment, subsets with three elements are also easy because we simply choose an element to *not* include. This yields the four subsets $\{b, c, d\}, \{a, c, d\}, \{a, b, d\}, \{a, b, c\}$. And there is only one subset $\{a, b, c, d\}$ with all the elements.
With regards to two sets, there is a way to ensure that we have found all pairs, and the method generalizes. The idea is to first pair a up with each of the other elements. Then we move on to b and pair it up with every element other than itself and a because a was already paired up with b. Finally, we pair c up with every element other than itself and a and b, which means only d. This yields the array

$$\begin{array}{ccc} \{a,b\} & \{a,c\} & \{a,d\} \\ & \{b,c\} & \{b,d\} \\ & & \{c,d\} \end{array}.$$

So the power set is

$$\{\varnothing, a, b, c, d, ab, ac, ad, bc, bd, cd, abc, abd, acd, bcd, abcd\},$$

where we have omitted set notation among the elements for the sake of having some breathing room.

Solution 3.4. It is clear that there is a Cartesian product involved, so we can use the independent multiplication principle to get the answer $2 \cdot 3 \cdot 4 = 24$.

Solution 3.5. We are asked to find the number of ways in which the entries of a k-tuple of 1's and -1's can be multiplied to equal to 1. For $k = 1$, the only tuple is (1) so the answer is 1. For $k \geq 2$, the idea is that first $k-1$ entries can be anything, and each possibility for the first $k-1$ entries leads to exactly one possible a_k. Formally, we use the map

$$(a_1, a_2, \ldots, a_k) \mapsto (a_1, a_2, \ldots, a_{k-1}),$$

which goes from the set of solutions to the Cartesian product $\{-1, 1\}^{k-1}$. This is a bijection because one value of a_k works and the other does not. Thus, the answer is

$$|\{-1, 1\}^{k-1}| = 2^{k-1}.$$

The method that we just used is a common one in mathematics: give full freedom to some parameters, and this results in the remaining parameters being fixed.

Solution 3.10. Let n be a positive integer and let $(A_1, A_2, \ldots, A_n)$ be a list of non-empty finite sets. For each index k such that $1 \leq k \leq n-1$, no matter how the first k indices of an n-tuple in $T = A_1 \times A_2 \times \cdots \times A_n$ are given a valid assignment of entries, any one of the $|A_{k+1}|$ elements of A_{k+1} can serve as a valid entry in the $(k+1)^{\text{th}}$ index and there are no other valid entries for the $(k+1)^{\text{th}}$ index. Thus, for each index k such that $2 \leq k \leq n$, the kth dependence number exists and is equal to $|A_k|$. Of course the first dependence number exists and is equal to $|A_1|$. Thus, T is symmetrically dependent.

Solution 3.14. We claim that the number of bijections from S to itself is the same as the number of permutations of S. To prove this, we will construct a bijection between the sets

$$B = \{b : b \text{ is a bijection from } S \text{ to } S\},$$
$$P = \{p : p \text{ is a permutation of } S\}.$$

Of course, B is non-empty because the identity function $\mathrm{Id}_S : s \mapsto s$ is a bijection from S to S, and P is known to be non-empty, so it makes sense to speak of a bijection from B to P.
The idea that we will pursue is that fixing a permutation (i.e. an ordering) of S and composing it with each of the elements of B produces each element of P once and produces only elements of P. We can make this precise as follows. Since $|S| = n$, there exists a bijection $f : [n] \to S$, which we take to be our fixed ordering of S. We define the operator (this is a term for a function whose domain is itself a set of functions) $T : B \to P$ by $T(b) = b \circ f$, and claim that T is bijective. Indeed, it is easy to verify that $b \circ f \in P$ and:

- T is injective because if $b_1, b_2 \in B$ such that $b_1 \circ f = b_2 \circ f$, then

$$b_1 = b_1 \circ (f \circ f^{-1}) = (b_1 \circ f) \circ f^{-1} = (b_2 \circ f) \circ f^{-1} = b_2 \circ (f \circ f^{-1}) = b_2.$$

- T is surjective because if $p \in P$, then we can reverse-engineer a function $b \in B$ such that $b \circ f = p$: just define $b = p \circ f^{-1}$ so that

$$b \circ f = (p \circ f^{-1}) \circ f = p \circ (f^{-1} \circ f) = p.$$

Thus, the operator T is a bijection, proving $B \approx P$. Since P is known to be finite with cardinality $n!$, the bijection principle implies that B is finite with cardinality

$$|B| = |P| = n!.$$

In some contexts, such as abstract algebra, permutations of S are defined to be these bijections $b : S \to S$, but we have instead used a definition of permutations that indexes the elements of S according to a section of $\mathbb{Z}_+$ so that instilling an order is the purpose of the permutation. In any case, we have just proven that the number of "permutations" of S is the same either way.

Solution 3.15. It is possible to express the solution using the bijection principle with a Cartesian product, but it not usually the case that solutions are written so formally. While our work has been relatively formal so far, we have to admit that, once it is clear that the foundations lie on solid ground, it would be debilitating and stifling to not write solutions to problems more casually. We will write this solution in a relaxed way and the interested reader can make it more formal.
Consider the m friends to be one mega-person. Then there are $n-m+1$ "people" who must be seated. The number of permutations of a set of $n-m+1$ people is $(n-m+1)!$. For each permutation of the $n-m+1$ "people," there are $m!$ permutations of the m friends within the mega-person. So there are $(n-m+1)! \cdot m!$ suitable permutations in total.

Solution 3.16. Each couple can be considered to be a mega-person that occupies two consecutive seats. Then there are $n!$ ways of seating the n mega-people among the n mega-seats. Since each couple can be seated in 2 ways, the final answer is $n! \cdot 2^n$ via a Cartesian product.

Solution 3.20. The reader may wish to have the statements of the bijection principle (Theorem 1.10) and division principle (Theorem 3.17) at hand so that it may be compared with the correspondence principle (Theorem 3.19).

1. Let X and Y be non-empty sets. First we will show that a function $f : X \to Y$ is a bijection if and only if it is a 1-to-1 correspondence. Indeed, f is a bijection if and only if it is an injection and a surjection; f is an injection if and only if the preimage of each $y \in Y$ has at most one element; and f is a surjection if and only if the preimage of each $y \in Y$ has at least one element. So f is a bijection if and only if the preimage of each $y \in Y$ is a singleton, which is true if and only if f is a 1-to-1 correspondence. From comparing the statement of the correspondence principle with the statement of the bijection principle, it requires no further reasoning to see that the latter is the restriction of the former to the $k = 1$ case.

2. Suppose A is a non-empty set and $P = \{A_1, A_2, \dots, A_n\}$ is a partition of A for a positive integer n, where each A_i is finite with
$$|A_1| = |A_2| = \cdots = |A_n| = k$$
for some positive integer k. We define $f : A \to P$ as the function that maps each $a \in A$ to the unique A_i that contains a. Then $f^{-1}(A_i) = A_i$ for each index $1 \leq i \leq n$, and so
$$|f^{-1}(A_i)| = |A_i| = k.$$
Thus, f is a k-to-1 correspondence, which allows us to conclude from the correspondence principle that A is finite and
$$n = |P| = \frac{|A|}{k}.$$

Solution 3.24. We know that there are $(n-1)!$ (which is an even integer for $n \geq 3$) circular permutations of the n keys. Each circular permutation has a "flipped" version that has the keys in the same order but with opposite orientation. For each such clockwise-counterclockwise pair, we map both elements to the key ring that has its keys in this order. This produces a 2-to-1 correspondence, so the the number of key rings is $\dfrac{(n-1)!}{2}$.

Solution 3.29. Imagine that we take out $k-1$ of the seats so that we can place them in between consecutive people later. This leaves $n-(k-1) = n-k+1$ seats and we can assign the k people to them in $\dbinom{n-k+1}{k}$

ways, and then we pop the seats that we took out back in to ensure that at least one seat is in between each pair of consecutive people. But this calculation assumed that the k people are not distinguishable. Since they are distinguishable and each of the $\binom{n-k+1}{k}$ ways allows for $k!$ permutations of the k people, resulting in a final count of

$$\binom{n-k+1}{k} \cdot k! = (n-k+1)(n-k)\cdots(n-2k+2).$$

In retrospect, this calculation actually makes sense as a permutation, where we assign k people to $n-k+1$ seats and pop back the extra $k-1$ seats with 1 of them in between each pair of consecutive seated people.

Solution 3.30. If we identify a rectangle in the grid, then extending the sides of the rectangle yields an unordered pair of distinct horizontal grid segments and an unordered pair of distinct vertical grid segments. In the other direction, if we have an unordered pair of distinct horizontal grid segments and an unordered pair of distinct vertical grid segments, then the intersection of the space between the horizontal ones and the space between the vertical ones produces a rectangle in the grid. This is a bijection. However, note that there are actually $v+1$ horizontal grid lines and $h+1$ vertical grid lines, so the answer is $\binom{h+1}{2} \cdot \binom{v+1}{2}$.

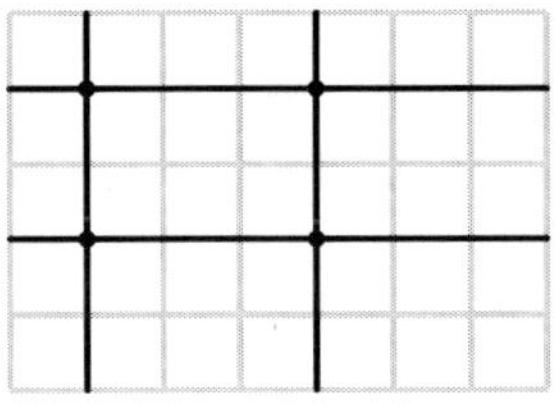

Solution 3.31. The set of integers c such that $p \leq c \leq q$ is $S = \{p, p+1, p+2, \ldots, q-1, q\}$, which has cardinality $q-p+1$. Let C_k be the set of k-combinations of S. A list $(c_1, c_2, \ldots, c_k)$ such that $p \leq c_1 < c_2 < \cdots < c_k \leq q$ has all of the c_i distinct, and so it can be mapped to the element of C_k that has the same elements. We claim that this map from L to C_k is a bijection. Indeed, since each element of C_k contains k distinct elements, they can be uniquely ordered from least to highest to produce an element of L. Thus, the preimage of each element of C_k is a singleton, proving the bijectivity of our map. The set L is finite with cardinality equal to the number of k-combinations of S. which is $\binom{q-p+1}{k}$.

Solution 3.38. This is the number of permutations of the multinomial set $f : \{A, B\} \to \mathbb{Z}_{\geq 0}$, defined by $f(A) = a$ and $f(B) = b$. Thus, the answer is

$$\frac{n!}{a! \cdot b!} = \binom{a+b}{a, b} = \binom{a+b}{b} = \binom{a+b}{a}.$$

Solution 3.39. There is a bijection between suitable sequences of the g_i and h_j, and the number of ways of permuting n identical copies of the letter g with m identical copies of the letter h. The reason is that the g_i must appear in ascending order and the h_j must appear in ascending order, and any permutation of the g_i and h_j is suitable if the g_i are ascending and h_j as ascending. Thus, the answer is

$$\frac{(n+m)!}{n! \cdot m!} = \binom{n+m}{n} = \binom{n+m}{m}.$$

See Example 4.19 for an application of the counting technique used this problem.

Solution 4.9. The cardinality of the sample space is $(2 \cdot 2)^n = 2^{2n}$ because at each of the n steps, there are 2 directions in which A can go and 2 directions in which B can go. Moreover, the two do not affect each other, nor is any direction affected by past choices.

Determining the cardinality of the set of successes is more interesting. One method is to do casework on the points at which A and B can coincide, after the completion of step n. The points that A or B can occupy after step n are those whose x-coordinate and y-coordinate add up to n, because that is precisely what indicates that n steps have been taken in this problem. There are $n+1$ such points:

$$\{(0,n),(1,n-1),(2,n-2),\ldots,(n-1,1),(n,0)\}.$$

There are $\binom{n}{i}$ ways in which each of A or B can reach $(i,n-i)$ for $i=0,1,2,\ldots,n-1,n$. Moreover, the way in which A reaches $(i,n-i)$ does not affect how B reaches $(i,n-i)$, so the total number of possibilities is

$$\binom{n}{0}\cdot\binom{n}{n}+\binom{n}{1}\cdot\binom{n}{n-1}+\cdots+\binom{n}{n-1}\cdot\binom{n}{1}+\binom{n}{n}\cdot\binom{n}{0}.$$

We could stop here, but a closed form is more desirable. A cleverer method is to recognize that, if A and B meet, then joining the two paths and reversing B's steps produces an increasing path from $(0,0)$ to (n,n). This map is bijective, so the answer is $\binom{2n}{n}$. As a bonus, we can equate the two expressions found to get the combinatorial identity

$$\sum_{i=0}^{n}\binom{n}{i}^2=\sum_{i=0}^{n}\binom{n}{i}\cdot\binom{n}{n-i}=\binom{2n}{n}$$

holds. The final probability is $\frac{1}{2^{2n}}\cdot\binom{2n}{n}$.

The reader might be interested in thinking about what happens if B starts at (m,n) for $m>n$ or $m<n$.

Solution 4.14. In accordance with the current theme, we will prove these identities by showing that it is possible to interpret them in terms of committee-building scenarios.

1. We have n people, out of which we want to select a committee of k people, and out of the k-person committee we want to select a sub-committee of m people. Directly, there are $\binom{n}{k}\binom{k}{m}$ ways to do this. Alternatively, we could select m people for the sub-committee out of the n people, and build the rest of the committee around the sub-committee. This method yields

$$\binom{n}{m}\binom{n-m}{k-m}$$

ways. Then we set the two computations equal to each other.

2. Setting $m=1$ in the first identity yields

$$\binom{n}{k}k=n\binom{n-1}{k-1},$$

which is equivalent to what we want to see.

3. We use a variation of the scenario designed for the first identity. The set of ways to have a sub-committee of m people can vary according to the size of the committee within which the sub-committee lies. The committee can have cardinality $k=m,m+1,\ldots,n$, and so we get the left side

$$\sum_{k=m}^{n}\binom{n}{k}\binom{k}{m}.$$

Alternatively, there are $\binom{n}{m}$ ways of selecting the sub-committee and then each of the remaining $n-m$ might or might not be in the rest of the larger committee. So the left side is equal to $\binom{n}{m}2^{n-m}$.

Solution 4.16. We will count the number of permutations of $\{1, 2, 3, \ldots, n+1\}$ (in this case, by permutation, we mean a bijection from the set to itself) that are not the identity permutation. This means at least one element will not map to itself. In fact, we will use the double counting method, so that we count the possibilities in two ways and can set the two expressions equal to each other. One method says that there are $(n+1)!$ permutations in total and we subtract 1 to eliminate the identity permutation, yielding $(n+1)! - 1$.
The other method uses a technique reminiscent of the proof of the hockey stick identity (Theorem 4.15): We know that some integer in $\{1, 2, 3, \ldots, n+1\}$ will not map to itself, so the well-ordering principle says that there is a minimum such integer in the set. Let this be k, where we know that $1 \leq k \leq n$. Note that k cannot be $n+1$ because that would imply that the first n values map to themselves, which would force $n+1$ to also map it itself in the bijection, which would create the identity permutation. In each case where $1 \leq k \leq n$, it means that the first $k-1$ values map to themselves, and k has $n+1-k$ values to which it can map (these exclude the first $k-1$ values and k itself). The remaining $n+1-k$ input elements can bijectively map to the remaining $n+1-k$ output elements in $(n+1-k)!$ ways. Since the cases are disjoint, we can sum this over $k = 1, 2, 3, \ldots, n$ to yield

$$1 \cdot 1! + 2 \cdot 2! + 3 \cdot 3! + \cdots + n \cdot n!$$

for the second count.

Solution 4.17. By applying the symmetry of Pascal's triangle and the hockey stick identity, we get

$$\begin{aligned}\sum_{i=0}^{n} \binom{k+i}{i} &= \binom{k}{0} + \binom{k+1}{1} + \cdots + \binom{k+n}{n} \\ &= \binom{k}{k} + \binom{k+1}{k} + \cdots + \binom{k+n}{k} = \sum_{i=k}^{k+n} \binom{i}{k} = \binom{k+n+1}{k+1} = \binom{k+n+1}{n}.\end{aligned}$$

Visually, this is just a reflection of the hockey stick identity across the central vertical line in Pascal's triangle.

Solution 4.20. There are no indices less than the first index $i = 1$, so $f(1) = 1$ holds vacuously. For each positive integer n, we can interpret the recurrence relation

$$n! = \sum_{k=1}^{n} f(k)(n-k)!$$

as double counting the number of permutations of $[n]$. The left side of the equation is the standard formula for counting permutations. The right side is trickier: it does casework on the smallest integer k for which we can say that the the first k entries $(a_1, a_2, \ldots, a_k)$ (of the permutation $(a_1, a_2, \ldots, a_n)$) is a permutation of $[k]$. The minimality of k means $(a_1, a_2, \ldots, a_k)$ is an indecomposable permutation of $[k]$, of which there are $f(k)$. Note that such a unique index k must exist for each permutation of $[n]$ by the well-ordering principle because $k = n$ works, even if all lower indices fail to have this property. The remaining entries $(a_{k+1}, a_{k+2}, \ldots, a_n)$ can be permuted in $(n-k)!$ ways. The formula follows from summing $f(k)(n-k)!$ over the disjoint cases $k = 1, 2, \ldots, n$. Using $f(1) = 1$ and the recurrence formula

$$f(n) = n! - \sum_{k=1}^{n-1} f(k)(n-k)!,$$

we can compute $(f(n))_{n=1}^{7} = (1, 1, 3, 13, 71, 461, 3447)$.

Solution 5.1. Computing the expansions yields:

$$\begin{aligned}(x+y)^0 &= 1, \\ (x+y)^1 &= x + y, \\ (x+y)^2 &= x^2 + 2xy + y^2, \\ (x+y)^3 &= x^3 + 3x^2y + 3xy^2 + y^3, \\ (x+y)^4 &= x^4 + 4x^3y + 6x^2y^2 + 4xy^3 + y^4.\end{aligned}$$

For each n, if the terms are placed in descending (or ascending) order of the power of x in the term, the coefficients seem to form the nth row of Pascal's triangle.

Solution 5.4. An idea is to use the binomial theorem and the symmetry of binomial coefficients to get

$$\begin{aligned}\left(x+\frac{1}{x}\right)^n &= \sum_{k=0}^{n}\binom{n}{k}x^{n-k}\left(\frac{1}{x}\right)^k = \sum_{k=0}^{n}\binom{n}{k}x^{n-2k}\\ &= \binom{n}{0}x^n+\binom{n}{1}x^{n-2}+\cdots+\binom{n}{n-1}\frac{1}{x^{n-2}}+\binom{n}{n}\frac{1}{x^n}\\ &= \binom{n}{0}x^n+\binom{n}{1}x^{n-2}+\cdots+\binom{n}{1}\frac{1}{x^{n-2}}+\binom{n}{0}\frac{1}{x^n}\\ &= \binom{n}{0}\left(x^n+\frac{1}{x^n}\right)+\binom{n}{1}\left(x^{n-2}+\frac{1}{x^{n-2}}\right)+\cdots,\end{aligned}$$

where the final term in the last line is a constant that pairs up with no other term if and only if n is even. Formally, we would use induction with the base cases $t = x + \dfrac{1}{x}$ and

$$x^2+\frac{1}{x^2} = \left(x+\frac{1}{x}\right)^2 - 2 = t^2 - 2.$$

In the inductive step, we can use the above manipulations to isolate

$$\binom{n}{0}\left(x^n+\frac{1}{x^n}\right) = x^n + \frac{1}{x^n}$$

in terms of $x^k + \dfrac{1}{x^k}$ for certain $k < n$.

An alternative recursive method that is simpler is to note that

$$tq_n(x) = \left(x+\frac{1}{x}\right)\cdot\left(x^n+\frac{1}{x^n}\right) = \left(x^{n+1}+\frac{1}{x^{n+1}}\right)+\left(x^{n-1}+\frac{1}{x^{n-1}}\right) = q_{n+1}(x)+q_{n-1}(x),$$

and so $q_{n+1}(x) = tq_n(x) - q_{n-1}(x)$. A linear recursion of polynomials yields another polynomial. Note that we again require $q_1(x) = t$ and $q_2(x) = t^2 - 2$ as the base cases.

Solution 5.11. We are seeking the coefficient of x^3 in the expansion of

$$(1+x)(1+x)(1+x+x^2)(1+x+x^2+x^3).$$

Note that we have a $+1$ term in each factor to account for the possibility of zero balls of the corresponding colour being contributed, since 1 is another way of writing x^0. The expansion is

$$1+4x+8x^2+11x^3+11x^4+8x^5+4x^6+x^7,$$

so the answer is 11.

Solution 5.12. We are seeking the sum of the coefficients of x^4 and x^5 in the expansion of

$$(x+x^2+x^3+x^4)^3 = x^3(1+x+x^2+x^3)^3,$$

which are the coefficients of x and x^2 in $(1+x+x^2+x^3)^3$. We could expand this completely, but there is a shorter approach. Since we are seeking the coefficients of x and x^2, terms of higher degree are redundant. So we first compute

$$(1+x+x^2)^2 = 1+2x+3x^2+2x^3+x^4.$$

Then we omit $2x^3+x^4$ and compute the product

$$(1+2x+3x^2)(1+x+x^2) = 1+3x+6x^2+5x^3+3x^4.$$

Therefore, the answer is $3+6=9$.

Solution 5.13. We will prove this by a generating function. For each index i, the binomial theorem tells us that $\binom{bt-bi}{t}$ is the coefficient of x^t in $(1+x)^{bt-bi}$. By more applications of the binomial theorem, the sum on the left side of the desired identity is the coefficient of x^t in

$$\begin{aligned}\sum_{i=0}^{t}(-1)^i\binom{t}{i}(1+x)^{bt-bi} &= \sum_{i=0}^{t}(-1)^i\binom{t}{i}\left((1+x)^b\right)^{t-i}\\ &= \left((1+x)^b-1\right)^t\\ &= \left(\binom{b}{1}x+\binom{b}{2}x^2+\cdots+\binom{b}{b}x^b\right)^t\\ &= x^t\cdot\left(\binom{b}{1}+\binom{b}{2}x+\cdots+\binom{b}{b}x^{b-1}\right)^t.\end{aligned}$$

So, the coefficient x^t is $\binom{b}{1}^t=b^t$, which is the right side of the stated identity.

Solution 5.18. This equality is proven by the following transformation between generating functions:

$$\prod_{k=0}^{\infty}\frac{1}{(1-x^{3k+1})(1-x^{3k+2})}=\frac{\prod_{k=1}^{\infty}(1-x^{3k})}{\prod_{k=1}^{\infty}(1-x^k)}=\prod_{k=1}^{\infty}(1+x^k+x^{2k}).$$

Here, we have used the difference of cubes factorization. By the same manipulations, except with the difference of mth powers factorization, we find that the number of ways of partitioning n into non-multiples of $m\geq 1$ is the number of ways of partitioning n so that no part appears m times or more.

Solution 5.19. Let S be the set of squares of positive integers. By the formula for a finite geometric series, the generating function of the partitions of the first kind is

$$1\cdot(1+x^2)\cdot(1+x^3+x^6)\cdot(1+x^4+x^8+x^{12})\cdots=\prod_{m=1}^{\infty}\left(\sum_{k=0}^{m-1}x^{km}\right)=\prod_{m=1}^{\infty}\frac{1-x^{m^2}}{1-x^m}.$$

By telescoping all elements of the numerator with some elements of the denominator, this is equal to

$$\prod_{m\in\mathbb{Z}_+\setminus S}\frac{1}{1-x^m},$$

which is a closed form of the generating function for the partitions of the second kind.

Solution 5.21. We will proceed by induction on integers $n\geq 5$. In the base case, we get

$$\binom{10}{5}=252<256=4^4.$$

For the inductive step, we find that

$$\frac{\binom{2(n+1)}{n+1}}{\binom{2n}{n}}=\frac{\left(\frac{(2n+2)!}{(n+1)!(n+1)!}\right)}{\left(\frac{(2n)!}{n!n!}\right)}=\frac{(2n+1)(2n+2)}{(n+1)^2}=\frac{2(2n+1)}{n+1}<\frac{2(2n+2)}{n+1}=4.$$

Therefore, we can go from the induction hypothesis

$$\binom{2n}{n}<4^{n-1}$$

to

$$\binom{2(n+1)}{n+1} < 4 \cdot \binom{2n}{n} < 4 \cdot 4^{n-1} = 4^n,$$

which completes the induction.

Solution 5.22. As advised, we compute the binomial expansions

$$\begin{aligned}
(1+1)^n &= \binom{n}{0} + \binom{n}{1} + \binom{n}{2} + \binom{n}{3} + \cdots, \\
(1+i)^n &= \binom{n}{0} + i\binom{n}{1} - \binom{n}{2} - i\binom{n}{3} + \cdots, \\
(1-1)^n &= \binom{n}{0} - \binom{n}{1} + \binom{n}{2} - \binom{n}{3} + \cdots, \\
(1-i)^n &= \binom{n}{0} - i\binom{n}{1} - \binom{n}{2} + i\binom{n}{3} + \cdots.
\end{aligned}$$

When these four equations are added, we can see that the binomial coefficients with bottom entry $1, 2, 3$ disappear due to vertical cancellation. Because of the cyclic patterns in the powers of $1, i, -1, -i$, the same cancellation occurs whenever the bottom entry is not divisible by 4, leaving us with

$$4 \cdot \left[\binom{n}{0} + \binom{n}{4} + \binom{n}{8} + \cdots\right].$$

Of course, $(1+1)^n = 2^n$ and $(1-1)^n = 0$. By de Moivre's formula,

$$\begin{aligned}
(1+i)^n &= \sqrt{2}^n \left(\cos\frac{\pi}{4} + i\sin\frac{\pi}{4}\right)^n = \sqrt{2}^n \left(\cos\frac{n\pi}{4} + i\sin\frac{n\pi}{4}\right), \\
(1-i)^n &= \sqrt{2}^n \left(\cos\left(-\frac{\pi}{4}\right) + i\sin\left(-\frac{\pi}{4}\right)\right)^n = \sqrt{2}^n \left(\cos\frac{n\pi}{4} - i\sin\frac{n\pi}{4}\right).
\end{aligned}$$

Therefore,

$$(1+1)^n + (1+i)^n + (1-1)^n + (1-i)^n = 2^n + 2\sqrt{2}^n \cdot \cos\frac{n\pi}{4},$$

which yields the identity

$$\sum_{k=0}^{\infty} \binom{n}{4k} = 2^{n-2} + 2^{\frac{n-2}{2}} \cdot \cos\frac{n\pi}{4}.$$

Solution 5.24. We will be using the expression for multinomial coefficients in terms of factorials. Letting n, m and the k_i be as stated, the sum is equal to

$$\sum_{i=1}^{m} k_i \cdot \frac{(n-1)!}{k_1! \cdot k_2! \cdots k_m!} = \frac{(n-1)!}{k_1! \cdot k_2! \cdots k_m!} \cdot \sum_{i=1}^{m} k_i = \frac{n \cdot (n-1)!}{k_1! \cdot k_2! \cdots k_m!} = \binom{n}{k_1, k_2, \ldots, k_m}.$$

For Pascal's identity, let n and k be positive integers such that $n \geq k+1$. By choosing $m = 2$ and $k_1 = k \geq 1$ and $k_2 = n - k \geq 1$, this multinomial identity becomes

$$\binom{n-1}{k-1, n-k} + \binom{n-1}{k, n-k-1} = \binom{n}{k, n-k}.$$

By translating these multinomial coefficients into binomial coefficients, we get

$$\binom{n-1}{k-1} + \binom{n-1}{k} = \binom{n}{k},$$

which is Pascal's identity.

Solution 5.26. There exists a proof in [4] that divides both sides by 2^n and interprets the left side as the sum of the probabilities of a partition of a sample space, but we will show a mechanical inductive proof instead to match the theme of the section. For each non-negative integer n, let

$$p(n) = \sum_{k=0}^{n} \binom{n+k}{k} \cdot \frac{1}{2^k}.$$

It is clear that $p(0) = 1 = 2^0$, so the base case holds. Now, suppose $p(n) = 2^n$ for some non-negative integer n. We wish to show that $p(n+1) = 2^{n+1}$. By applying Pascal's identity to each term of $p(n+1)$ after the first term, it turns out that

$$\begin{aligned} p(n+1) &= 1 + \sum_{k=1}^{n+1} \binom{n+1+k}{k} \cdot \frac{1}{2^k} \\ &= 1 + \sum_{k=1}^{n+1} \left[\binom{n+k}{k-1} + \binom{n+k}{k}\right] \cdot \frac{1}{2^k} \\ &= 1 + \sum_{k=0}^{n} \binom{n+k+1}{k} \cdot \frac{1}{2^{k+1}} + \sum_{k=1}^{n+1} \binom{n+k}{k} \cdot \frac{1}{2^k} \\ &= 1 + \frac{1}{2} \cdot \left(p(n+1) - \binom{2n+2}{n+1} \cdot \frac{1}{2^{n+1}}\right) + \left(p(n) - 1 + \binom{2n+1}{n+1} \cdot \frac{1}{2^{n+1}}\right). \end{aligned}$$

Via further manipulations and using the symmetry of Pascal's triangle at the end, we get

$$\begin{aligned} \frac{1}{2}p(n+1) &= p(n) + \binom{2n+1}{n+1} \cdot \frac{1}{2^{n+1}} - \binom{2n+2}{n+1} \cdot \frac{1}{2^{n+2}} \\ &= p(n) + \frac{1}{2^{n+1}} \cdot \left(\binom{2n+1}{n+1} - \binom{2n+2}{n+1} \cdot \frac{1}{2}\right) \\ &= p(n) + \frac{1}{2^{n+1}} \cdot \left(\binom{2n+1}{n+1} - \frac{2n+2}{n+1} \cdot \binom{2n+1}{n} \cdot \frac{1}{2}\right) \\ &= p(n) + \frac{1}{2^{n+1}} \cdot \left(\binom{2n+1}{n+1} - \binom{2n+1}{n}\right) = p(n). \end{aligned}$$

By the induction hypothesis, $p(n+1) = 2 \cdot p(n) = 2^{n+1}$.

Solution 5.27. We will make repeated use of the identity

$$\binom{n}{k} = \frac{n}{k}\binom{n-1}{k-1}$$

for integers $n \geq k \geq 1$.

1. The identity is clearly true for $n = 0$, so we may assume that n is positive. Then

$$\begin{aligned} \sum_{k=0}^{n} k\binom{n}{k} = \sum_{k=1}^{n} k\binom{n}{k} &= \sum_{k=1}^{n} n\binom{n-1}{k-1} \\ &= n \cdot \sum_{k=1}^{n} \binom{n-1}{k-1} = n \cdot \sum_{k=0}^{n-1} \binom{n-1}{k} = n \cdot 2^{n-1}. \end{aligned}$$

For the corollary, since $[n]$ has n elements, it can have subsets of cardinality $k = 0, 1, 2 \ldots, n$. For each such k, there are $\binom{n}{k}$ distinct subsets of cardinality k. Therefore, the sum of cardinalities is

$$\sum_{k=0}^{n} k\binom{n}{k} = n \cdot 2^{n-1}.$$

2. If n is positive, then we can compute

$$\begin{aligned}\sum_{k=0}^{n} k^2\binom{n}{k} &= \sum_{k=1}^{n} k^2\binom{n}{k} = \sum_{k=1}^{n} kn\binom{n-1}{k-1}\\ &= n\cdot\sum_{k=1}^{n} k\binom{n-1}{k-1} = n\cdot\sum_{k=0}^{n-1}(k+1)\binom{n-1}{k}\\ &= n\cdot\sum_{k=0}^{n-1} k\binom{n-1}{k} + n\cdot\sum_{k=0}^{n-1}\binom{n-1}{k}.\end{aligned}$$

By the first part and the sum of a row of Pascal's triangle, this is equal to

$$n\cdot(n-1)2^{n-2} + n\cdot 2^{n-1} = \left[\frac{n(n-1)}{2} + n\right]\cdot 2^{n-1} = \frac{n(n+1)}{2}\cdot 2^{n-1} = \binom{n+1}{2}\cdot 2^{n-1}.$$

3. For all non-negative integers n and the sum of a row of Pascal's triangle,

$$\sum_{k=0}^{n}\frac{1}{k+1}\binom{n}{k} = \sum_{k=0}^{n}\frac{1}{n+1}\binom{n+1}{k+1} = \frac{1}{n+1}\sum_{k=0}^{n}\binom{n+1}{k+1} = \frac{2^{n+1}-1}{n+1}.$$

Solution 5.29. The reader may wish to review Vandermonde's identity (Theorem 5.9) as we will be applying it.

1. We append a null term to the bottom of the sum and null terms to the top as follows:

$$\sum_{k=1}^{m}\binom{m}{k}\binom{n-1}{k-1} = \sum_{k=1}^{m}\binom{m}{k}\binom{n-1}{n-k} = \sum_{k=0}^{n}\binom{m}{k}\binom{n-1}{n-k}.$$

Note that we applied the symmetry of Pascal's triangle in the first step. By Vandermonde's identity, this is equal to $\binom{m+n-1}{n}$.

2. For positive integers n,

$$\sum_{k=0}^{n} k\binom{n}{k}^2 = \sum_{k=1}^{n} n\binom{n-1}{k-1}\binom{n}{k} = n\cdot\sum_{k=1}^{n}\binom{n-1}{k-1}\binom{n}{k}.$$

By the first part, this is equal to $n\cdot\binom{2n-1}{n}$.

3. By the definition of Narayana numbers, we compute

$$\sum_{k=1}^{n} N(n,k) = \sum_{k=1}^{n}\frac{1}{n}\binom{n}{k}\binom{n}{k-1} = \frac{1}{n}\cdot\sum_{k=1}^{n}\binom{n}{k}\binom{n}{k-1}.$$

By the first part, this is equal to

$$\frac{1}{n}\binom{2n}{n+1} = \frac{1}{n}\cdot\frac{(2n)!}{(n+1)!(n-1)!} = \frac{1}{n+1}\cdot\frac{(2n)!}{n!\cdot n!} = \frac{1}{n+1}\binom{2n}{n},$$

which is the nth Catalan number.

Solution 6.5. Let n_i be an element of $[n]$; we will later specify i to be 1 or 2 depending on whether n_i is odd or even, respectively. For each $a \in A$, let $c(a)$ denote the sum of the 1's and (-1)'s contributed by a to $\sum_{k=1}^{n_i} (-1)^{k+1} S_k$. Let t be the number of A_k in which a appears. We will split up the evaluation of $c(a)$ into two cases: t satisfies $1 \leq t \leq n_i$ or $n_i < t \leq n$. If $1 \leq t \leq n_i$, then

$$c(a) = \sum_{k=1}^{t} (-1)^{k+1} \binom{t}{k},$$

where the upper bound on the index k gets cut off after t because a does not lie in the intersection of more than t sets. This evaluates to

$$1 - \sum_{k=0}^{t} (-1)^k \binom{t}{k} = 1 - (-1+1)^t = 1.$$

If $n_i < t \leq n$, then $c(a) = \sum_{k=1}^{n_i} (-1)^{k+1} \binom{t}{k}$, where the upper bound on the index k gets cut off after n_i because higher order terms from PIE are not included in Bonferroni. By Example 5.25, this evaluates to

$$1 - \sum_{k=0}^{n_i} (-1)^k \binom{t}{k} = 1 - (-1)^{n_i} \binom{t-1}{n_i}$$

because the sum is an alternating sum of an initial segment of a row of Pascal's triangle. Doing casework on the parity of n_i, we get

$$c(a) = \begin{cases} 1 - \dbinom{t-1}{n_i} \leq 1 - 1 = 0 < 1 & \text{if } i = 2, \\ 1 + \dbinom{t-1}{n_i} \geq 1 + 1 = 2 > 1 & \text{if } i = 1. \end{cases}$$

Combining our results for $1 \leq t \leq n_i$ and $n_i < t \leq n$ yields

$$\sum_{k=1}^{n_2} (-1)^{k+1} S_k = \sum_{a \in A} c(a) \leq \sum_{a \in A} 1 = |A|,$$
$$\sum_{k=1}^{n_1} (-1)^{k+1} S_k = \sum_{a \in A} c(a) \geq \sum_{a \in A} 1 = |A|,$$

which is what we wanted to prove. Note that we cannot necessarily claim that these inequalities are strict because it might be true that there are no elements of A that lie in more than n_i of the A_k.

Solution 6.6. Since the stated formula for $\max(x_i)_{i=1}^n$ is symmetric in the x_i, we may assume without loss of generality that $x_1 \leq x_2 \leq \cdots \leq x_n$, as the proof of the identity for this ordering will prove the formula for all other orderings. The left side is equal to x_n. For each integer $\ell \in [n]$, we will count how many times x_ℓ appears as a summand on the right side; to be clear, if there is a tie for a minimum element in one of the summands (since the x_i are only non-decreasing and not necessarily strictly increasing), the winner is the element with the lowest index among the ties. For $\ell = n$, x_ℓ is not the minimum of any k-tuples $(x_j)_{j \in J}$ if $k \geq 2$; so x_n appears only once on the right side, specifically for $k = 1$. If $\ell \in [n-1]$, then for each integer $k \in [n]$, the number of k-tuples $(x_j)_{j \in J}$ in which x_ℓ is the minimum element is $\binom{n-\ell}{k-1}$. Backtracking a bit, we actually need k to satisfy $0 \leq k-1 \leq n-\ell$, because only $n-\ell$ of the x_i have index greater than ℓ. For k such that $k-1 > n-\ell$, we do not have to worry about x_ℓ being the minimum of a k-tuple $(x_j)_{j \in J}$. Thus, if $\ell \in [n-1]$, then the signed count of the $(\pm x_\ell)$'s that appear as a summand on the right side is

$$\sum_{k=1}^{n-\ell+1} (-1)^{k+1} \binom{n-\ell}{k-1} = \sum_{k=0}^{n-\ell} (-1)^k \binom{n-\ell}{k} = (-1+1)^{n-\ell} = 0.$$

This proves the first identity, and the second one follows because

$$\min(x_i)_{i=1}^n = -\max(-x_i)_{i=1}^n = -\sum_{k=1}^{n}(-1)^{k+1}\sum_{\substack{J\subseteq[n]\\|J|=k}}\min(-x_j)_{j\in J}$$
$$= \sum_{k=1}^{n}(-1)^{k+1}\sum_{\substack{J\subseteq[n]\\|J|=k}}(-\min(-x_j)_{j\in J}) = \sum_{k=1}^{n}(-1)^{k+1}\sum_{\substack{J\subseteq[n]\\|J|=k}}\max(x_j)_{j\in J}.$$

After some rearrangement, the formulas for $n = 2, 3$ are

$$\min(x_1, x_2) + \max(x_1, x_2) = x_1 + x_2,$$
$$\max(x_1, x_2, x_3) + \min(x_1, x_2) + \min(x_2, x_3) + \min(x_1, x_3) = x_1 + x_2 + x_3 + \min(x_1, x_2, x_3),$$
$$\min(x_1, x_2, x_3) + \max(x_1, x_2) + \max(x_2, x_3) + \max(x_1, x_3) = x_1 + x_2 + x_3 + \max(x_1, x_2, x_3).$$

The identities involving the gcd and lcm functions follow from the formulas for these functions in terms of the prime factorizations of a, b, c (see Volume 3). We simply have to do casework on each prime p that divides at least one of a, b, c and let the x_i be the multiplicities of p across a, b, c.

Solution 6.10. Let Id_s be the arithmetic function defined by $\mathrm{Id}_s(n) = n^s$ for every positive integer s. By the Möbius inversion formula, $S_{J_s} = \mathrm{Id}_s$ if and only if

$$(\mu * \mathrm{Id}_s)(n) = J_s(n).$$

It is easy to deduce that J_s is multiplicative from the formula for J_s. Moreover,

$$(\mu * \mathrm{Id}_s)(1) = \mu(1)\mathrm{Id}_s(1) = 1 = J_s(1).$$

Thus, we only have to verify that

$$(\mu * \mathrm{Id}_s)(p^k) = J_s(p^k)$$

for all primes p and positive integers k. Fixing p and k, the definition of the Möbius function μ yields

$$(\mu * \mathrm{Id}_s)(p^k) = \sum_{i=0}^{k}\mu(p^i)\mathrm{Id}_s(p^{k-i}) = \mu(1)\cdot\mathrm{Id}_s(p^k) + \mu(p)\cdot\mathrm{Id}_s(p^{k-1})$$
$$= 1\cdot p^{ks} + (-1)\cdot p^{(k-1)s} = (p^k)^s\left(1 - \frac{1}{p^s}\right) = J_s(p^k),$$

which completes the proof.

Solution 6.15. For any positive integer m, the number of potential edges in a graph with m is vertices is $\binom{m}{2}$, so the total number of graphs on m vertices is $2^{\binom{m}{2}}$. Let the given vertices be labelled $1, 2, \ldots, n$, which is possible since they are distinguishable. For each $k \in [n]$, let G_k be the set of graphs on these n vertices such that vertex k is isolated. By complementary counting and the symmetric variant of PIE, the answer is

$$2^{\binom{n}{2}} - \left|\bigcup_{k=1}^{n} G_k\right| = 2^{\binom{n}{2}} - \sum_{k=1}^{n}(-1)^{k+1}\sum_{\substack{J\subseteq[n]\\|J|=k}}\left|\bigcap_{j\in J} G_j\right|$$
$$= 2^{\binom{n}{2}} + \sum_{k=1}^{n}(-1)^{k}\sum_{\substack{J\subseteq[n]\\|J|=k}}2^{\binom{n-k}{2}}$$
$$= 2^{\binom{n}{2}} + \sum_{k=1}^{n}(-1)^{k}\binom{n}{k}2^{\binom{n-k}{2}}$$
$$= \sum_{k=0}^{n}(-1)^{k}\binom{n}{k}2^{\binom{n-k}{2}}.$$

As usual, we absorbed the initial $2^{\binom{n}{2}}$ term into the sum as the 0th term in the final step.

Solution 7.2. The first ball could go into one of k boxes. Then the second ball could go into one of $k-1$ boxes. We continue in this way until the nth ball can go into one of $k-n+1$ boxes. Thus, the total number of possibilities is

$$k(k-1)\cdots(k-n+1) = \frac{k!}{(k-n)!}.$$

This is simply a permutation.

Solution 7.3. Since $k \geq n$, some n of the k boxes will have exactly one ball and the other $k-n$ boxes will have no balls. So, this is simply a matter of choosing n of the k boxes, which gives an answer of $\binom{k}{n}$.

Solution 7.5. For each of the k distinguishable boxes, there are n tubs from which a ball can be chosen to be put into the box. Since there are k balls in each tub, we will not run out of balls in any tub in this way. By the multiplication principle, the answer is $\underbrace{n \cdot n \cdots n}_{k \text{ copies of } n} = n^k$.

Solution 7.11. There is a bijection between increasing functions $f : [n] \to [m]$ and n-subsets of $[m]$. This is because each n-subset of $[m]$ (which corresponds to the range of a function f) can be permuted in exactly one increasing order. Thus, the number of increasing functions $f : [n] \to [m]$ is $\binom{m}{n}$.

Finding the number of non-decreasing functions is a bit trickier. Since non-decreasing functions can have several inputs mapping to the same output, our correspondence this time will take the multinomial set corresponding to the range of a non-decreasing function $g : [n] \to [m]$, with each element of $[m]$ appearing as many times as an element of $[n]$ maps to it. This is a bijection because each such multinomial set chosen from $[m]$ (where the sum of multiplicities is n) leads to exactly one non-decreasing permutation. Thus, the number of non-decreasing functions $g : [n] \to [m]$ is the multiset coefficient

$$\left(\!\!\binom{m}{n}\!\!\right) = \binom{n+m-1}{m-1}.$$

Solution 7.12. We are seeking the number of m-tuples

$$(k_1, k_2, \ldots, k_m)$$

where the k_i are non-negative integers satisfying

$$k_1 + k_2 + \cdots + k_m = n.$$

So, we want to know the number of weak m-compositions of n. The answer is the multiset coefficient

$$\binom{n+m-1}{m-1} = \left(\!\!\binom{m}{n}\!\!\right).$$

Solution 7.13. A tuple $(a_1, a_2, \ldots, a_k)$ is a solution if and only if, for each $i \in [m]$,

$$\nu_{p_i}(a_1) + \nu_{p_i}(a_2) + \cdots + \nu_{p_i}(a_k) = \nu_{p_i}(a_1 a_2 \cdots a_k) = \nu_{p_i}(n) = e_i.$$

So, we are counting a Cartesian product of weak k-compositions, and the answer is

$$\binom{e_1+k-1}{k-1} \cdot \binom{e_2+k-1}{k-1} \cdots \binom{e_m+k-1}{k-1}.$$

If the a_j are allowed to be negative, then we find the number of positive solutions as above, and then we arbitrarily assign signs to $a_1, a_2, \ldots, a_{k-1}$, which fixes the sign of a_k since the sign of n is fixed (as positive). Thus, allowing negatives results in multiplying the above formula by 2^{k-1}.

Solution 7.15. We do not need the formula for the κ function, though it is possible to use it in an algebraic proof of this result. Instead, we observe that the upper bound $(m-1)k$ on the index n is the highest possible sum of this k-composition, since each of the k components is at most $m-1$, and the lower bound 0 on n is the lowest achievable sum. By the strong multiplication principle on the k components of the k-composition, each of which can be anything in $\{0, 1, 2, \ldots, m-1\}$, we get the cardinality

$$|\{0, 1, 2, \ldots, m-1\}^k| = m^k.$$

Solution 7.17. By the definition of multiset coefficients and Pascal's identity,

$$\left(\!\!\binom{n}{k-1}\!\!\right) + \left(\!\!\binom{n-1}{k}\!\!\right) = \binom{(k-1)+n-1}{n-1} + \binom{k+(n-1)-1}{(n-1)-1}$$
$$= \binom{k+n-2}{n-1} + \binom{k+n-2}{n-2} = \binom{k+n-1}{n-1} = \left(\!\!\binom{n}{k}\!\!\right).$$

A combinatorial interpretation of this is that the right side $\left(\!\!\binom{n}{k}\!\!\right)$ is the number of ways of distributing k indistinguishable balls across n distinguishable boxes, whereas the left side splits it up into the case where a specific box gets at least one ball and the case where that specific box gets no balls at all.

Solution 7.19. In much the same way that we started the solution to Example 7.18, the solution set is in bijection with the Cartesian product $P \times C$ where P is the set of permutations of the n distinguishable balls, and C is the set of ways of distributing n indistinguishable balls to k distinguishable boxes such that empty boxes are allowed. This gives an answer of

$$|P \times C| = |P| \cdot |C| = n!\binom{n+k-1}{k-1} = \frac{(n+k-1)!}{(k-1)!}.$$

This makes sense because another way of thinking about it is that we are permuting n distinguishable balls and $k-1$ indistinguishable sticks. There would be $(n+k-1)!$ permutations if the sticks were distinguishable too, and then we have to divide out by the $(k-1)!$-fold symmetry from the sticks' indistinguishability.

Solution 7.20. The c_i must be chosen from among the integers $p, p+1, p+2, \ldots, q$. For each integer i such that $p \leq i \leq q$, let a_i be the number of times that i appears as an element in the list $(c_1, c_2, \ldots, c_k)$. Then the list $(a_p, a_{p+1}, \ldots, a_q)$ of non-negative integers satisfies

$$a_p + a_{p+1} + \cdots + a_q = k,$$

making it a weak $(q-p+1)$-composition of k. This map is injective because, given the list $(a_p, a_{p+1}, \ldots, a_q)$, we can go backwards to recover $(c_1, c_2, \ldots, c_k)$ by having a_i copies of each i such that $p \leq i \leq q$. Moreover, the map is surjective because we can go backwards like this with every list of non-negative integers $(a_p, a_{p+1}, \ldots, a_q)$ such that $a_p + a_{p+1} + \cdots + a_q = k$. So there is a bijection and it is equivalent to count the number of weak $(q-p+1)$-compositions of k, which we know to number

$$\binom{k+(q-p+1)-1}{(q-p+1)-1} = \binom{k+q-p}{q-p}.$$

Solution 7.24. Let $S = \{a_1, a_2, \ldots, a_n\}$ be a set of n elements.

- The only 1-partition of S is $\{S\}$, so $\left\{ {n \atop 1} \right\} = 1$.

- The only n-partition of S is the collection of singletons of elements of S, so $\left\{ {n \atop n} \right\} = 1$.

- A 2-partition is a set of two sets $\{A, B\}$ such that $A \sqcup B = S$. So, there are two choices of sets in which each element of S can be placed, either A or B, giving 2^n distributions. However, this double-counts the number of possibilities because $\{A, S \backslash A\}$ is the same partition as $\{S \backslash A, A\}$. We divide by 2 to get 2^{n-1}. Finally, we remove $\{S, \varnothing\}$ from the possibilities to get a final answer of $\left\{ {n \atop 2} \right\} = 2^{n-1} - 1$.

- As we will prove, the only way to have an $(n-1)$-partition of S is to have one set of two elements and $n-2$ singletons. If there is a set with more than two elements, then even if the rest of the elements are split into singletons, there will be fewer than $n-1$ sets in the partition. So all sets in the partition have one or two elements. If there are two or more sets with two elements, then we run into the same issue, so there is at most one set with two elements. Finally, if all of the sets are singletons, then there are more than $n-1$ sets in the partition. Thus, it is equivalent to count the number of two-element subsets of S, which gives an answer of $\left\{ {n \atop n-1} \right\} = \binom{n}{2}$.

Solution 7.25. Recall that $\left\{ {n+1 \atop k+1} \right\}$ denotes the number of $(k+1)$-partitions of a set S of $n+1$ elements. Fix a particular element a of S. Now we will split the possibilities for a into two cases: in a $(k+1)$-partition, either a lives in a singleton by itself, or a is in one of the $k+1$ subsets in the $(k+1)$-partition that must contain other elements of S. In the former case, there are $\left\{ {n \atop k} \right\}$ partitions because the remaining n elements have to be partitioned into k subsets. In the latter case, we first remove a from consideration and the remaining n elements are partitioned into $k+1$ subsets in $\left\{ {n \atop k+1} \right\}$ ways, and then a is placed into one of these $k+1$ subsets. This produces the desired identity.

Solution 7.31. We compute the desired expressions below.

- The only 1-partition of n is $\langle n \rangle$, so $\left| {n \atop 1} \right| = 1$.

- The only n-partition of n is $\langle \underbrace{1, 1, \ldots, 1}_{n \text{ copies of } 1} \rangle$, so $\left| {n \atop n} \right| = 1$.

- We place n dots in a row and consider the number of ways a divider can be placed in between two consecutive dots to produce a partition. Since there is symmetry across the middle, so we have to do something akin to dividing by 2 in order to not overcount. For odd n, we get $\frac{n-1}{2}$ partitions, and, for even n, we get $\frac{n-2}{2} + 1 = \frac{n}{2}$ partitions because the central divider does not produce a second partition. These two formulas can be unified using the floor function as $\left\lfloor \frac{n}{2} \right\rfloor$.

- Suppose we have a partition of n into $n-1$ parts. If there is an element equal to 3 or more, then this leaves $n-3$ and even if $n-3$ is split into 1's, it produces only $n-2$ elements, including the 3. So all elements are 1 or 2. There cannot be only 1's because that would exceed the number of prescribed parts. And there cannot be more than one element equal to 2 because it would cause the same contradiction as with having an element equal to 3 or more. So there is exactly one element equal to 2, and the rest are 1's. Thus, $\left| {n \atop n-1} \right| = 1$.

Solution 7.32. We know that $\left|\begin{matrix} n \\ k \end{matrix}\right|$ is the number of k-partitions of n. We split this into two cases: partitions where there is at least one element equal to 1, and partitions where all elements are strictly greater than 1. In the former case, there is a bijection with $(k-1)$-partitions of $n-1$ because we can remove 1 element that is known to equal to 1, and then we can partition the remaining $n-1$ into $k-1$ positive integers. So this case contributes $\left|\begin{matrix} n-1 \\ k-1 \end{matrix}\right|$ partitions. In the latter case, there is a bijection with k-partitions of $n-k$ because we can remove 1 from each element and still have a partition, as a result of each element being greater than 1. Bijectively, this case contributes $\left|\begin{matrix} n-k \\ k \end{matrix}\right|$ partitions.

Solution 8.3. We can count the number of unordered pairs of vertices, which is $\binom{n}{2}$.

Solution 8.8. Suppose an Eulerian circuit exists in a graph. For each vertex, every time the circuit passes through the vertex, there is an edge going towards the vertex and then another vertex going away from the vertex. So, every non-terminal vertex (as in, not where we choose to begin and end the traversal of the circuit) has an even number of edges attached to it. If we begin traversing the circuit on vertex v, then we have to end on it, so there is an even number of edges attached to v, plus 2 more (the first edge and last edge of the traversal), for a total even degree.

Solution 8.11. We prove the results in succession, as the first part helps with proving the second part:

1. We prove the result by induction. If $n = 1$, then there is no edge and the result holds. Now suppose the result holds for some $n \geq 1$ and that we are given a graph with $n+1$ vertices. Let $\mathcal{V}$ be the vertex set of the graph and $\mathcal{E}$ be the edge set of the graph. If all the vertices of this graph have degree greater than or equal to 2, then the handshaking lemma tells us that

$$2|\mathcal{E}| = \sum_{v \in \mathcal{V}} \deg v \geq 2(n+1).$$

   So $|\mathcal{E}| \geq n+1$ and the result follows. Otherwise, suppose there is a leaf. Then we can remove the leaf and the edge emanating from it to produce another connected graph with n vertices. By the induction hypothesis, this has at least $n-1$ edges, and so the original graph has at least n edges.

2. For contradiction, suppose there is a connected graph with exactly $n \geq 1$ vertices and exactly $n-1$ edges, yet it is not a tree. By the definition of a tree, this non-tree graph has a cycle. By Lemma 8.10, we can remove any edge from the cycle to produce a new graph that is still connected, but has n vertices and $n-2$ edges. This contradicts the first part of the problem.

Solution 8.13. The result is obviously true for a graph on a single vertex, so we can assume that $n \geq 2$. For each list of positive integers $(d_1, d_2, \ldots, d_n)$ such that

$$\sum_{i=1}^{n} d_i = 2n-2,$$

let $T(d_1, d_2, \ldots, d_n)$ be the set of distinct trees with the vertex set $[n] = \{1, 2, \ldots, n\}$, such that the vertex i has degree d_i for each integer $1 \leq i \leq n$. Since the collection of trees with the vertex set $[n]$ is the disjoint union of all the $T(d_1, d_2, \ldots, d_n)$, Theorem 8.12 tells us that we want to compute the number

$$\sum_{\substack{d_1+d_2+\cdots+d_n=2n-2 \\ d_i \text{ positive}}} |T(d_1, d_2, \ldots, d_n)| = \sum_{\substack{d_1+d_2+\cdots+d_n=2n-2 \\ d_i \text{ positive}}} \binom{n-2}{d_1-1, d_2-1, \ldots, d_n-1}.$$

Recall that the multinomial theorem (Theorem 5.6) says that for any positive integer ℓ and any non-negative integer m,

$$(x_1 + x_2 + \cdots + x_\ell)^m = \sum_{\substack{k_1+k_2+\cdots+k_\ell=m \\ k_i \text{ non-negative}}} \binom{m}{k_1, k_2, \ldots, k_\ell} x_1^{k_1} x_2^{k_2} \cdots x_\ell^{k_\ell}.$$

By substituting $x_1 = x_2 = \cdots = x_\ell = 1$, the sum of the multinomial coefficients is

$$\sum_{\substack{k_1+k_2+\cdots+k_\ell=m \\ k_i \text{ non-negative}}} \binom{m}{k_1, k_2, \ldots, k_\ell} = \ell^m.$$

Choosing $m = n - 2$ and $\ell = n$ yields

$$n^{n-2} = \sum_{\substack{k_1+k_2+\cdots+k_n=n-2 \\ k_i \text{ non-negative}}} \binom{n-2}{k_1, k_2, \ldots, k_n} = \sum_{\substack{d_1+d_2+\cdots+d_n=2n-2 \\ d_i \text{ positive}}} \binom{n-2}{d_1 - 1, d_2 - 1, \ldots, d_n - 1}.$$

Solution 8.20. Suppose for contradiction that every vertex has degree greater than or equal to 6. Let $\mathcal{V}$ be the set of vertices, V be the number of vertices, and E be the number of edges. Then the handshaking lemma (Theorem 8.5) tells us that

$$2E = \sum_{v \in \mathcal{V}} \deg v \geq 6V$$

or $E \geq 3V$. Our planarity criterion says that $E \leq 3V - 6$, which contradicts the previous statement. Thus, there must exist a vertex of degree less than or equal to 5.
As a side note, the result holds even if the planar graph is not connected. This is because every graph splits into connected subgraphs called components, so we can apply the result to each component. This was not a part of the problem, since we have not treated connected components in our exposition of graphs.

Solution 8.22. The idea is to look at the sum $\sum_{vw \in \mathcal{E}} (\deg v + \deg w)$. Firstly, since every vertex can be connected to at most one of v or w (otherwise, there would be a triangle), we use the weak upper bound $\deg v + \deg w \leq |\mathcal{V}|$, so the sum is at most $|\mathcal{E}| \cdot |\mathcal{V}|$. Secondly, $\deg v$ appears in $\deg v$ many summands because the sum iterates over all edges and v appears in $\deg v$ edges. As a result,

$$|\mathcal{E}| \cdot |\mathcal{V}| \geq \sum_{vw \in \mathcal{E}} (\deg v + \deg w) = \sum_{v \in \mathcal{V}} (\deg v \cdot \deg v) = \sum_{v \in \mathcal{V}} (\deg v)^2.$$

By a special case of the Cauchy-Schwarz inequality and the handshaking lemma, we get

$$|\mathcal{E}| \cdot |\mathcal{V}| \geq \sum_{v \in \mathcal{V}} (\deg v)^2 \geq \frac{1}{|\mathcal{V}|} \left(\sum_{v \in \mathcal{V}} \deg v \right)^2 = \frac{1}{|\mathcal{V}|} \cdot (2|\mathcal{E}|)^2,$$

which simplifies to $|\mathcal{E}| \leq \dfrac{|\mathcal{V}|^2}{4}$, as desired.

Solution 8.25. We will prove the results in sequence, each leading to the next.

1. Suppose the vertex set of the bipartite graph splits into the sets $\mathcal{V}_1$ and $\mathcal{V}_2$. Without loss of generality, suppose a cycle starts with a vertex in $\mathcal{V}_1$. Then the next vertex must be in $\mathcal{V}_2$ and we keep toggling between vertices in $\mathcal{V}_1$ and $\mathcal{V}_2$ in this way. Thus, if there is an odd number of edges in the cycle, then the first vertex and the final vertex cannot both be in $\mathcal{V}_1$, which contradicts the fact that cycles start and end with the same vertex. Thus, the cycle cannot have an odd number of edges.

2. A cycle must have at least 3 edges. However, 3 is odd, so by the last part, a cycle in a bipartite graph must have at least 4 edges. So bipartite graphs are triangle-free.

3. It is easy to see that $K_{3,3}$ is a connected graph with $V = 6$ vertices and $E = 9$ edges. Moreover, it is bipartite, so it is triangle-free. Then $E = 9 > 2 \cdot 6 - 4 = 2V - 4$, which means $K_{3,3}$ is not planar, by Theorem 8.23.

Solution 8.28. First note that every vertex has degree greater than or equal to 3, as neither a vertex of degree 1 nor a vertex of degree 2 is compatible with a convex polyhedron. Letting the vertex set be $\mathcal{V}$, the handshaking lemma (Theorem 8.5) asserts $2E = \sum_{v \in \mathcal{V}} \deg v \geq 3V$.

Solution 8.34. As with the icosahedron, we first subtract the number of edges from the number of unordered pairs of vertices to find the total number of diagonals. This is $\binom{20}{2} - 30 = 160$. Now we have to subtract the number of face diagonals. There are 12 pentagonal faces with $\binom{5}{2} - 5 = 5$ face diagonals each. So the final answer is $160 - 5 \cdot 12 = 100$.

Solution 8.38. Let the first colour be red and the second colour be blue. Suppose $R(r, b)$ exists. Let the edges of the complete graph $K_{R(r,b)}$ be coloured arbitrarily. Then there exists a corresponding "inverse" colouring that swaps red edges for blue edges and blue edges for red edges. This new graph still has $R(r, b)$ vertices, and, by the existence and definition of $R(r, b)$, it has a red r-clique or a blue b-clique. Inverting back, the original arbitrarily coloured graph must have a blue r-clique and a red b-clique. Thus, $R(b, r)$ must exist and satisfy $R(b, r) \leq R(r, b)$. If we start with the existence of $R(b, r)$, the same argument shows that $R(r, b)$ exists and satisfies $R(r, b) \leq R(b, r)$. So one exists if and only if the other does. In the case that they exist, combining $R(b, r) \leq R(r, b)$ and $R(r, b) \leq R(b, r)$ with antisymmetry yields that $R(r, b) = R(b, r)$.

Solution 8.39. By the symmetry result (Problem 8.38), it suffices to only prove $R(2, t) = t$. Let the first colour be red and the second colour be blue. Firstly, it is true that $R(2, t) \leq t$ because if the complete graph K_t is not coloured as a blue t-clique, then there must be at least one red edge, which is the same as there being a red 2-clique. Secondly, if $R(2, t) < t$, then there exists the colouring that colours all edges with blue, yet there is neither a blue t-clique nor a red edge. Thus, the equality $R(2, t) = t$ holds.

Solution 8.45. To prove that $R(r, b) > (r - 1)(b - 1)$, it suffices to show that there exists a red-blue edge colouring of the complete graph on $(r - 1)(b - 1)$ vertices such that there is no red r-clique and that there is no blue b-clique in it. To this end, we partition the vertices into $b - 1$ groupings, each with $r - 1$ vertices. Within each separate grouping of $r - 1$ vertices, every pair of vertices is connected by a red edge. Afterwards, every pair of vertices that occupy two different groupings is attached by a blue edge. This completes the colouring. It is clear that, while each grouping is a red $(r - 1)$-clique, there exists no red r-clique. As for blue cliques, a blue clique cannot have more than one element from the same grouping, as elements within the same grouping are connected by the same edge; so, the number of vertices in a blue clique is at most the number of the groupings, which is $b - 1$. Therefore, no red r-clique exists and no blue b-clique exists.

Solution 8.46. Let the colours be $[c] = \{1, 2, \ldots, c\}$. In one direction, suppose $R(r_1, r_2, \ldots, r_c)$ exists. Let the edges of the complete graph $K_{R(r_1,r_2,\ldots,r_c)}$ be coloured arbitrarily. Then, for each $i \in [c]$, we alter the colouring so that all i-coloured edges are replaced by $\sigma^{-1}(i)$-coloured edges. This new graph still has $R(r_1, r_2, \ldots, r_c)$ vertices, so there exists a colour $j \in [c]$ such that the newly coloured graph has an j-colored r_j-clique. Let $\sigma^{-1}(j) = i$ so that $\sigma(i) = j$. So there is a j-coloured $r_{\sigma(i)}$-clique in the newly coloured graph. This means that the original arbitrarily coloured graph must have an i-coloured $r_{\sigma(i)}$-clique. Thus,

$$R(r_{\sigma(1)}, r_{\sigma(2)}, \ldots, r_{\sigma(c)})$$

must exist and satisfy

$$R(r_{\sigma(1)}, r_{\sigma(2)}, \ldots, r_{\sigma(c)}) \leq R(r_1, r_2, \ldots, r_c).$$

If we start with the existence of

$$R(r_{\sigma(1)}, r_{\sigma(2)}, \ldots, r_{\sigma(c)}),$$

then applying the bijection σ^{-1} to what we have already established shows that $R(r_1, r_2, \ldots, r_c)$ exists and satisfies

$$R(r_1, r_2, \ldots, r_c) \leq R(r_{\sigma(1)}, r_{\sigma(2)}, \ldots, r_{\sigma(c)}).$$

So one exists if and only if the other does. In the case that one (and so both) exist, combining the two inequalities with antisymmetry yields

$$R(r_1, r_2, \ldots, r_c) = R(r_{\sigma(1)}, r_{\sigma(2)}, \ldots, r_{\sigma(c)}).$$

Solution 8.47. Let the colours be $[c] = \{1, 2, \ldots, c\}$. Firstly, it is true that $R(\underbrace{2, \ldots, 2}_{c-1 \text{ of 2's}}, t) \leq t$ because if the complete graph K_t is not coloured as a c-coloured t-clique, then there must be at least one edge of one of the first $c-1$ colours, which is the same as there being a 2-clique of one of the first $c-1$ colours. Secondly, if we suppose $R(\underbrace{2, \ldots, 2}_{c-1 \text{ of 2's}}, t) < t$ for contradiction, then we can construct the colouring that colours all edges with the final colour c, yet there is neither a c-coloured t-clique nor an edge of one of the other $c-1$ colours. Thus, the equality $R(\underbrace{2, \ldots, 2}_{c-1 \text{ of 2's}}, t) = t$ holds.

Solution 8.55. The essence of the first two parts is that the deletion-contraction recurrence allows us to recursively prove the existence of the chromatic polynomial by performing strong induction on the number of edges of a graph with a fixed number of vertices $n \geq 1$.

1. By complementary counting, $\chi_G(k)$ is the number of vertex colourings of G where the only edge both of whose vertices are *allowed* to have the same colour is uv, minus those vertex colourings of G where u and v *do* have the same colour. The former, meaning the universal set of this application of the subtraction principle, is in bijection with proper vertex colourings of $G\backslash uv$ in the natural way: map each such colouring of G to the same colouring of $G\backslash uv$. The latter, meaning the colourings of G where u and v share a colour, are in bijection with G/uv, also in a natural way: map each such colouring of G to the colouring of G/uv where every vertex other than u, v preserves its colour and the new vertex w inherits the common colour of u and v. With some thought, it becomes that these maps are indeed bijections. Thus, the deletion-contraction recurrence holds:

$$\chi_G(k) = \chi_{G\backslash uv}(k) - \chi_{G/uv}(k).$$

2. Let n be the number of vertices of G. Clearly, if there are k colours and n vertices with no edges, then k^n proper vertex colourings are possible by the independent multiplication principle (Theorem 3.1), since each vertex is unhindered by the others. This forms the base case for strong induction. Now suppose there exists an integer $E \geq 0$ such that any graph with n vertices and E or fewer edges has a chromatic function that is a polynomial. Let H be a graph with n vertices and $E + 1 \geq 1$ edges, with uv being a particular edge. Then $H\backslash uv$ and H/uv each have at most E edges, so the strong induction hypothesis tells us that

$$\chi_{G\backslash uv}(k) \text{ and } \chi_{G/uv}(k)$$

are both polynomials in k. By the deletion-contraction recurrence,

$$\chi_G(k) = \chi_{G\backslash uv}(k) - \chi_{G/uv}(k)$$

is also a polynomial in k.

3. (a) If $k < n$, then the pigeonhole principle says that some two vertices will share a colour, which is a contradiction because the complete graph K_n on n vertices has an edge between every pair of vertices. So, it makes sense that $0, 1, 2, \ldots, k-1$ are all roots of the stated polynomial. On the other hand, if $k \geq n$, then the stated polynomial follows from the dependent multiplication principle (Theorem 3.9), as each vertex has one fewer colour to choose its colour from, compared to the previous vertex. Differently stated, we are looking at an n-permutation of $[k]$.

(b) Once we know the formula, it is easy to proceed by induction on $n \geq 1$ by removing a leaf: In the base case, there are k possible colourings of a single vertex. Supposing the result holds for any tree on $n \geq 1$ vertices, let T be a tree with $n+1$ vertices. By Theorem 8.9, there must be a leaf in T. Removing the leaf and its attached edge yields a tree with n vertices, whose chromatic polynomial must be $k(k-1)^{n-1}$ by the induction hypothesis. Attaching back the removed leaf, there are $k-1$ ways to colour it, as only the colour of its parent is not allowed. This yields a chromatic polynomial of

$$k(k-1)^{n-1} \cdot (k-1) = k(k-1)^n,$$

which completes the induction.

(c) Let the vertex set be $\{v_1, v_2, \ldots, v_n\}$, where the edges are $v_i v_{i+1}$ for each $i \in [n]$ (we take v_{n+1} to be v_1). For each set $i \in [n]$, let N_i denote the number of (not necessarily proper) colourings of C_n such that vertex v_i has the same colour as vertex v_{i+1}. By the principle of inclusion-exclusion (Theorem 6.1) and complementary counting,

$$\chi_{C_n}(k) = k^n - \left|\bigcup_{i=1}^{n} N_i\right| = k^n - \sum_{i=1}^{n} (-1)^{i+1} \sum_{\substack{S \subseteq [n] \\ |S|=i}} \left|\bigcap_{j \in S} N_j\right|.$$

If $i < n$, then each inner summand $\left|\bigcap_{j \in S} N_j\right|$ equal k^{n-i} because i of the vertices have their colours fixed and the other $n-i$ vertices have full freedom over their colours. On the other hand, if $i = n$, then the inner summand is k, since all vertices have the same one colour. So, by the symmetric principle of inclusion-exclusion (Corollary 6.2), we find that

$$\begin{aligned}
\chi_{C_n}(k) &= k^n - \left[\sum_{i=1}^{n-1} (-1)^{i+1} \binom{n}{i} k^{n-i}\right] - (-1)^{n+1} \binom{n}{n} k \\
&= (-1)^0 \binom{n}{0} k^{n-0} + \left[\sum_{i=1}^{n-1} (-1)^i \binom{n}{i} k^{n-i}\right] + (-1)^n k \\
&= \left[\sum_{i=0}^{n-1} (-1)^i \binom{n}{i} k^{n-i}\right] + (-1)^n k \\
&= \left[\sum_{i=0}^{n} (-1)^i \binom{n}{i} k^{n-i}\right] - (-1)^n \binom{n}{n} k^{n-n} + (-1)^n k \\
&= (k-1)^n + (-1)^n (k-1),
\end{aligned}$$

where we used the binomial theorem at the end.

Solution 9.8. Let n and L be as stated.

1. For one direction, suppose L is pairwise disjoint. For any integer k such that $2 \leq k \leq n$ and any k indices $1 \leq i_1 < i_2 < \cdots < i_k \leq n$, it holds that

$$A_{i_1} \cap A_{i_2} \cap \cdots \cap A_{i_k} \subseteq A_{i_1} \cap A_{i_2} = \varnothing.$$

Since the only subset of $\varnothing$ is $\varnothing$, we get that $A_{i_1} \cap A_{i_2} \cap \cdots \cap A_{i_k} = \varnothing$ and so L is mutually disjoint. The converse is true because being pairwise disjoint is merely the special case of $k = 2$ in the definition of being mutually disjoint.

2. For one direction, suppose $A_i \cap A_j$ occurs almost never for every pair of indices i, j such that $1 \leq i < j \leq n$. Let k be any integer such that $2 \leq k \leq n$ and let $1 \leq i_1 < i_2 < \cdots < i_k \leq n$ be any k indices. By the first Kolmogorov axiom and monotonicity,

$$0 \leq \mathbb{P}(A_{i_1} \cap A_{i_2} \cap \cdots \cap A_{i_k}) \leq \mathbb{P}(A_{i_1} \cap A_{i_2}) = 0.$$

By antisymmetry,

$$\mathbb{P}(A_{i_1} \cap A_{i_2} \cap \cdots \cap A_{i_k}) = 0,$$

which completes this direction of the proof. The converse holds simply by taking $k = 2$.

3. Suppose L is mutually exclusive. Then, for every pair of indices i, j such that $1 \le i < j \le n$, it holds that $A_i \cap A_j = \varnothing$. Subsequently, $\mathbb{P}(A_i \cap A_j) = 0$, which proves that L is almost mutually exclusive.

4. Suppose L is collectively exhaustive. Then

$$A_1 \cup A_2 \cup \cdots \cup A_n = \Omega,$$

and so

$$\mathbb{P}(A_1 \cup A_2 \cup \cdots \cup A_n) = 1,$$

proving that L is almost collectively exhaustive.

5. Suppose our finite probability space is positive.
   - Suppose L is almost mutually exclusive. Then knowing $\mathbb{P}(A_i \cap A_j) = 0$ allows us to conclude that $A_i \cap A_j = \varnothing$, for every pair of indices i, j such that $1 \le i < j \le n$. Thus, L is mutually exclusive.
   - Suppose L is almost collectively exhaustive. Then $B = A_1 \cup A_2 \cup \cdots \cup A_n$ satisfies $\mathbb{P}(B) = 1$. By complementary probability, $\mathbb{P}(\overline{B}) = 0$, and so $\overline{B} = \varnothing$. Therefore, $B = \Omega$, proving that L is collectively exhaustive.

Solution 9.10. By monotonicity and the upper bound on probabilities, if $\mathbb{P}(A \cup B) = \mathbb{P}(A) \cdot \mathbb{P}(B)$ then

$$\mathbb{P}(A) \le \mathbb{P}(A \cup B) = \mathbb{P}(A) \cdot \mathbb{P}(B) \le \mathbb{P}(A),$$
$$\mathbb{P}(B) \le \mathbb{P}(A \cup B) = \mathbb{P}(A) \cdot \mathbb{P}(B) \le \mathbb{P}(B).$$

Since each sequence is sandwiched between equal upper and lower bounds, antisymmetry causes all inequalities to be flattened into equalities, which yields

$$\mathbb{P}(A) = \mathbb{P}(B) = \mathbb{P}(A) \cdot \mathbb{P}(B) = \mathbb{P}(A \cup B).$$

Then $\mathbb{P}(A) = \mathbb{P}(A)^2$ and $\mathbb{P}(B) = \mathbb{P}(B)^2$. The only solutions to the equation

$$x^2 - x = x(x-1) = 0$$

are $x = 0$ or $x = 1$, so $\mathbb{P}(A) = \mathbb{P}(B)$ are both 0 or both 1.
Conversely, suppose $\mathbb{P}(A) = \mathbb{P}(B) = 0$ or $\mathbb{P}(A) = \mathbb{P}(B) = 1$. In the former case, the probabilistic principle of inclusion-exclusion for two sets tells us that

$$0 \le \mathbb{P}(A \cup B) = \mathbb{P}(A) + \mathbb{P}(B) - \mathbb{P}(A \cap B) = -\mathbb{P}(A \cap B) \le 0,$$

and so $\mathbb{P}(A \cup B) = 0 = \mathbb{P}(A) \cdot \mathbb{P}(B)$.
In the latter case, monotonicity and the upper bound on probabilities yield

$$1 = \mathbb{P}(A) \le \mathbb{P}(A \cup B) \le 1,$$

so $\mathbb{P}(A \cup B) = 1 = \mathbb{P}(A) \cdot \mathbb{P}(B)$.

Solution 9.13. Suppose $\mathbb{P}(A) \ne 0$ and $\mathbb{P}(B) \ne 0$. By the definition of conditional probability,

$$\mathbb{P}(A \mid B) = \mathbb{P}(B \mid A) \iff \frac{\mathbb{P}(A \cap B)}{\mathbb{P}(B)} = \frac{\mathbb{P}(B \cap A)}{\mathbb{P}(A)} \iff \mathbb{P}(A) = \mathbb{P}(B).$$

Solution 9.16. The second solution will use the first result:

1. A is independent of A if and only if $\mathbb{P}(A \cap A) = \mathbb{P}(A) \cdot \mathbb{P}(A)$, which is equivalent to $\mathbb{P}(A) = (\mathbb{P}(A))^2$. By solving the quadratic $x^2 - x = 0$, we see that this is true if and only if $\mathbb{P}(A) = 0$ or $\mathbb{P}(A) = 1$.

2. The idea is to start off with the simplifying assumptions that $A = \varnothing$ and $B = C$, and we will add other restrictions if needed later. Then
$$\begin{aligned} \mathbb{P}(A \cap B \cap C) &= \mathbb{P}(\varnothing \cap B \cap B) = \mathbb{P}(\varnothing) = 0, \\ \mathbb{P}(A) \cdot \mathbb{P}(B) \cdot \mathbb{P}(C) &= \mathbb{P}(\varnothing) \cdot \mathbb{P}(B) \cdot \mathbb{P}(B) = 0, \end{aligned}$$
equating which yields
$$\mathbb{P}(A \cap B \cap C) = \mathbb{P}(A) \cdot \mathbb{P}(B) \cdot \mathbb{P}(C).$$
We want one of the pairs to fail to be independent. The two pairs involving A will succeed, so our only chance of failure is for it to not hold that
$$\mathbb{P}(B \cap C) = \mathbb{P}(B) \cdot \mathbb{P}(C).$$
Since $B = C$, this equation holds if and only if B is self-independent, which we found to be true in the first part if and only if B occurs almost never or almost surely. So we just need a construction where B occurs neither almost never nor almost surely. Let $\Omega = \{x, y, z\}$ with $p(x) = 0$ and $p(y) = p(z) = \frac{1}{2}$. Choosing $B = \{y\}$, we are done.

Solution 9.17. If A and B are both independent and almost mutually exclusive, then
$$\begin{aligned} \mathbb{P}(A \cap B) &= 0, \\ \mathbb{P}(A \cap B) &= \mathbb{P}(A) \cdot \mathbb{P}(B). \end{aligned}$$
Equating them, we get $\mathbb{P}(A) \cdot \mathbb{P}(B) = 0$, which is equivalent to at least one of $\mathbb{P}(A)$ or $\mathbb{P}(B)$ being 0. Conversely, suppose at least one of $\mathbb{P}(A)$ or $\mathbb{P}(B)$ is 0. By monotonicity, $A \cap B \subseteq A$ and $A \cap B \subseteq B$ imply that
$$0 \leq \mathbb{P}(A \cap B) \leq \min\{\mathbb{P}(A), \mathbb{P}(B)\} = 0.$$
By antisymmetry, $\mathbb{P}(A \cap B) = 0 = \mathbb{P}(A) \cdot \mathbb{P}(B)$, proving that A and B are almost mutually exclusive and independent.

Solution 9.19. Let n and L be as stated.

1. Suppose L is mutually independent. Let $1 \leq i \leq n$ be any index and J be a subset of $[n] \backslash \{i\}$ such that $\mathbb{P}\left(\bigcap_{j \in J} A_j\right) \neq 0$. By the multiplicativity stemming from mutual independence,
$$\mathbb{P}\left(A_i \,\middle|\, \bigcap_{j \in J} A_j\right) = \frac{\mathbb{P}\left(A_i \cap \left(\bigcap_{j \in J} A_j\right)\right)}{\mathbb{P}\left(\bigcap_{j \in J} A_j\right)} = \frac{\mathbb{P}(A_i) \cdot \mathbb{P}\left(\bigcap_{j \in J} A_j\right)}{\mathbb{P}\left(\bigcap_{j \in J} A_j\right)} = \mathbb{P}(A_i).$$

2. Suppose the stated hypothesis. Let I be a non-empty subset of $[n]$. If I is a singleton, then the equation for mutual independence is automatically satisfied, so we may assume that $|I| = k \geq 2$. Let I be written as $\{i_1, i_2, \ldots, i_k\}$ where $i_1 < i_2 < \cdots < i_k$. By the chain rule for events,
$$\begin{aligned} \mathbb{P}\left(\bigcap_{i \in I} A_i\right) = \mathbb{P}\left(\bigcap_{j=1}^{k} A_{i_j}\right) &= \mathbb{P}(A_{i_1}) \cdot \prod_{j=2}^{k} \mathbb{P}\left(A_{i_j} \,\middle|\, \bigcap_{\ell=1}^{j-1} A_{i_\ell}\right) \\ &= \mathbb{P}(A_{i_1}) \cdot \prod_{j=2}^{k} \mathbb{P}(A_{i_j}) = \prod_{j=1}^{k} \mathbb{P}(A_{i_j}) = \prod_{i \in I} \mathbb{P}(A_i), \end{aligned}$$
where we used the hypothesis in each factor of the equation resulting from the chain rule. Note that the index i is unrelated to the placeholder symbol in the i_j or i_ℓ. We used the symbol i in both cases due to the limited available choices for symbols that are typically used for indexing purposes.

Solution 9.21. By the law of total probability,

$$\mathbb{P}(A) = \sum_{i=1}^{n} \mathbb{P}(A \cap B_i) = \sum_{i=1}^{n} \mathbb{P}(B_i \mid A) \cdot \mathbb{P}(A),$$

where the sum on the far right is the original part of this exercise. Dividing both sides by $\mathbb{P}(A)$ yields the desired identity. Since B and $\overline{B}$ form a partition of Ω, the corollary holds.

Solution 9.27. For each integer k such that $0 \leq k \leq n$, the number of k-element subsets of $[n]$ is $\binom{n}{k}$. The total number of subsets is 2^n. Letting f be the random variable that assigns to each subset of $[n]$ its cardinality, we get

$$\mathbb{E}(f) = \frac{1}{2^n} \sum_{k=0}^{n} k \binom{n}{k} = \frac{n2^{n-1}}{2^n} = \frac{n}{2}.$$

We used Problem 5.27 to evaluate the sum.

Solution 9.31. If $n = 1$, then the only bijection $\sigma : [1] \to [1]$ is the identity map and it has 1 fixed point, so the answer is 1. Now we assume that $n \geq 2$.
Let the sample space Ω be the set of bijections $\sigma : [n] \to [n]$ with the uniform probability on it. For each index $1 \leq i \leq n$, let f_i be the random variable that indicates whether i is a fixed point of σ. That is, we define $f_i : \Omega \to \{0, 1\}$ by

$$f_i(\sigma) = \begin{cases} 1 & \text{if } \sigma(i) = i \\ 0 & \text{if } \sigma(i) \neq i \end{cases}.$$

Then we are looking to compute

$$\mathbb{E}(f_1 + \cdots + f_n) = \mathbb{E}(f_1) + \cdots + \mathbb{E}(f_n),$$

where we have used the linearity of expectation. For any index i,

$$\mathbb{E}(f_i) = \sum_{\sigma \in \Omega} p(\sigma) f_i(\sigma) = \sum_{\sigma \in \Omega} \frac{1}{n!} \cdot f_i(\sigma) = \frac{1}{n!} \cdot \sum_{\sigma \in \Omega} f_i(\sigma) = \frac{1}{n!} \cdot |\{\sigma \in \Omega : \sigma(i) = i\}|.$$

So we need to find the cardinality of the set of bijections that fix i. It is easy to see that this set is in bijection with the set of bijections $\sigma : [n-1] \to [n-1]$, which has cardinality $(n-1)!$. Therefore, the answer is

$$\mathbb{E}\left(\sum_{i=1}^{n} f_i\right) = \sum_{i=1}^{n} \mathbb{E}(f_i) = \sum_{i=1}^{n} \frac{(n-1)!}{n!} = \sum_{i=1}^{n} \frac{1}{n} = n \cdot \frac{1}{n} = 1.$$

This matches our answer for $n = 1$, so the answer is strangely 1 for all positive integers n.

Solution 10.3. Based on the levels shown, the numbers in level n should sum to 3^n. This is due to the fact that, for each positive integer k, the sum of level $k+1$ consists of exactly 3 copies of the number assigned to each sphere in level k; this is due to the fact that each sphere is tangent to exactly 3 spheres in the level below it. Since the sum of level 0 is $1 = 3^0$, and we multiply the total by 3 every time we increase the level, it follows from an induction argument that that the sum of the numbers in level n is 3^n, for each non-negative integer n.

Solution 10.4. In a given partition of $[n+1]$, the element $n+1$ inhabits a set of $k+1$ elements in a partition for some $k \in \{0, 1, 2, \ldots, n\}$. There are $\binom{n}{k}$ ways to choose the other k elements of the set in which $n+1$ lies, and the remaining $(n+1) - (k+1) = n-k$ elements of $[n+1]$ can be partitioned in B_{n-k} ways. By casework and the symmetry of Pascal's triangle, we get

$$B_{n+1} = \sum_{k=0}^{n} \binom{n}{k} B_{n-k} = \sum_{k=0}^{n} \binom{n}{n-k} B_k = \sum_{k=0}^{n} \binom{n}{k} B_k.$$

Solution 10.7. All of the identities can be proven by induction on $n \geq 1$. The base case $n = 1$ is easy to verify for all of them, so we will only show the inductive steps. Suppose all of the identities hold for some positive integer n. Then the inductive steps are:

1. $\sum_{k=1}^{n+1} F_k = F_{n+1} + \sum_{k=1}^{n} F_k = F_{n+1} + F_{n+2} - 1 = F_{n+3} - 1 = F_{(n+1)+2} - 1$

2. $\sum_{k=0}^{n} F_{2k+1} = F_{2n+1} + \sum_{k=0}^{n-1} F_{2k+1} = F_{2n+1} + F_{2n} = F_{2n+2} = F_{2(n+1)}$

3. $\sum_{k=1}^{n+1} F_{2k} = F_{2n+2} + \sum_{k=1}^{n} F_{2k} = F_{2n+2} + F_{2n+1} - 1 = F_{2n+3} - 1 = F_{2(n+1)+1} - 1$

4. $\sum_{k=1}^{n+1} F_k^2 = F_{n+1}^2 + \sum_{k=1}^{n} F_k^2 = F_{n+1}^2 + F_n F_{n+1} = F_{n+1}(F_{n+1} + F_n) = F_{n+1}F_{n+2}$

Thus, all of the identities hold by induction.

Solution 10.9. Testing a few initial values, we find that t_2 can be reached in 1 way and t_3 can be reached in 2 ways. Technically, t_1 can also be reached in 1 way as well since we are already standing on it. It seems like t_n can be reached in F_n ways, where F_n is the nth Fibonacci number. Indeed, with these base cases of an induction proof established, we find that, for any positive integer $i \geq 3$, t_i can be reached in one way from t_{i-1} and another way from t_{i-2}. In a manner that is reminiscent of Pascal's method, there are

$$F_{i-1} + F_{i-2} = F_i$$

ways of reaching t_i, for each index $i \in [n]$. Taking $i = n$ yields the answer of F_n.

Solution 10.18. For each non-negative integer n, let T_n denote the number of rooted full binary trees on n parent nodes. As we know,

$$T_0 = 1, T_1 = 1, T_2 = 2, T_3 = 5.$$

We can show by strong induction on $n \geq 0$ that $T_n = C_n$, as follows. If we remove the root node of a rooted full binary tree on $n+1$ parent nodes along with the two edges emanating from the root, then we are left with two rooted full binary trees (a left tree and a right tree) with a total of n parent nodes. This removal is actually a bijection. The n remaining parent nodes can be distributed among the left tree and the right tree as $k + (n-k)$ for $k = 0, 1, 2, \ldots, n$. This leads to the recursion

$$T_{n+1} = \sum_{k=0}^{n} T_k T_{n-k}.$$

Since this is the same as the Catalan recursion and the base case $T_0 = 1 = C_0$ holds, we can complete the proof by strong induction.

Solution 11.3. We will prove by strong induction on $n \geq 0$ that there is only one admissible value of z_n. We are told that $z_0, z_1, \ldots, z_{k-1}$ are fixed, which takes care of the base case. Now, suppose $z_0, z_1, \ldots, z_n$ are fixed for some integer $n \geq k-1$. By the recursive relation give, it holds that

$$z_{n+1} = p(n+1) + \sum_{i=1}^{k} c_i z_{n+1-i},$$

The expression on the right side has only one value because, by the strong induction hypothesis, z_{n+1-i} has a unique value for $i = 1, 2, \ldots, k$. Thus, there is a unique value of the left side z_{n+1} as well, completing the induction.

Solution 11.5. Suppose there exists a constant b that can be added to both sides of the recurrence such that

$$\begin{aligned} b + a_n &= b + c + \sum_{i=1}^{k} c_i a_{n-i} = \sum_{i=1}^{k} c_i (b + a_{n-i}) \\ &= \sum_{i=1}^{k} c_i b + \sum_{i=1}^{k} c_i a_{n-i} = \sum_{i=1}^{k} c_i b + a_n - c. \end{aligned}$$

Simplifying and rearranging this equation gives

$$c = b(c_1 + c_2 + \cdots + c_k - 1).$$

If $c_1 + c_2 + \cdots + c_k = 1$, then $c = 0$, which contradicts the assumption that c is a non-zero constant. If $c_1 + c_2 + \cdots + c_k \neq 1$ then we can isolate

$$b = \frac{c}{c_1 + c_2 + \cdots + c_k - 1}.$$

Just to check, adding this b to to both sides of the recurrence relation yields

$$\begin{aligned} b + a_n &= b + c + \sum_{i=1}^{k} c_i a_{n-i} \\ &= \frac{c}{c_1 + c_2 + \cdots + c_k - 1} + c + \sum_{i=1}^{k} c_i a_{n-i} \\ &= \frac{c(c_1 + c_2 + \cdots + c_k)}{c_1 + c_2 + \cdots + c_k - 1} + \sum_{i=1}^{k} c_i a_{n-i} \\ &= \sum_{i=1}^{k} c_i \left(\frac{c}{c_1 + c_2 + \cdots + c_k - 1} + a_{n-i} \right) \\ &= \sum_{i=1}^{k} c_i (b + a_{n-i}). \end{aligned}$$

Thus, it suffices to find a formula for $(b_n)_{n=0}^{\infty} = (b + a_n)_{n=0}^{\infty}$ and subtract b from it. Note that the homogeneous linear recurrence for $(b_n)_{n=0}^{\infty}$ has depth k as well because $c_k \neq 0$.

Solution 11.7. For each integer n such that $n \geq k$, the discrete Fubini's principle tells us that

$$\begin{aligned} \sum_{i=1}^{k} c_i \gamma_{n-i} &= \sum_{i=1}^{k} c_i \sum_{j=1}^{t} \beta_j b_{j,n-i} = \sum_{i=1}^{k} \sum_{j=1}^{t} c_i \beta_j b_{j,n-i} \\ &= \sum_{j=1}^{t} \sum_{i=1}^{k} c_i \beta_j b_{j,n-i} = \sum_{j=1}^{t} \beta_j \sum_{i=1}^{k} c_i b_{j,n-i} \\ &= \sum_{j=1}^{t} \beta_j b_{j,n} = \gamma_n. \end{aligned}$$

Thus, $(\gamma_n)_{n=0}^{\infty}$ is a solution to this unrestricted recurrence relation.

Solution 11.11. The characteristic polynomial of the Lucas recurrence is $f(x) = x^2 - x - 1$, the roots of which are the golden ratio $\phi = \dfrac{1+\sqrt{5}}{2}$ and its radical conjugate $\psi = \dfrac{1-\sqrt{5}}{2}$. Solving the system

$$\begin{aligned} v_1 + v_2 &= 2, \\ v_1 \phi + v_2 \psi &= 1 \end{aligned}$$

yields $v_1 = 1$ and $v_2 = 1$. Therefore,

$$L_n = \phi^n + \psi^n$$

for all non-negative integers n.

Solution 11.12. The characteristic polynomial of the Pell recurrence is $f(x) = x^2 - 2x - 1$, the roots of which are

$$\alpha = 1 + \sqrt{2},$$
$$\beta = 1 - \sqrt{2}.$$

Solving the system

$$v_1 + v_2 = 0,$$
$$v_1\alpha + v_2\beta = 1$$

yields $v_1 = \frac{1}{2\sqrt{2}}$ and $v_2 = -\frac{1}{2\sqrt{2}}$. Therefore,

$$P_n = \frac{(1+\sqrt{2})^n - (1-\sqrt{2})^n}{2\sqrt{2}}$$

for all non-negative integers n.

Solution 11.14. By the formula for the Vandermonde determinant and a variation of the discrete Fubini's principle, the determinant of the given matrix is

$$\begin{aligned}\prod_{i=1}^{k-1}\prod_{j=i+1}^{k}((j-1)-(i-1)) = \prod_{i=1}^{k-1}\prod_{j=i+1}^{k}(j-i) = \prod_{j=2}^{k}\prod_{i=1}^{j-1}(j-i) = \prod_{j=2}^{k}(j-1)! = \prod_{j=1}^{k-1} j!\\ = 1!\cdot 2!\cdot 3!\cdots(k-1)! = 1^{k-1}\cdot 2^{k-2}\cdot 3^{k-3}\cdots(k-1)^1.\end{aligned}$$

This is the desired formula.

Solution 11.16. We see that, in the expansion of

$$A(x)C(x) = \left(\sum_{i=0}^{\infty} a_i x^i\right)\left(\sum_{i=1}^{k} c_i x^i\right),$$

the coefficient of x^n, for all positive integers n, is

$$a_0c_n + a_1c_{n-1} + a_2c_{n-2} + \cdots + a_{n-1}c_1,$$

where $c_j = 0$ for all integers $j > k$. Writing the terms in the opposite direction, we see that this is the recursion

$$\begin{aligned}a_n &= c_1a_{n-1} + c_2a_{n-2} + \cdots + c_{k-2}a_{n-k+2} + c_{k-1}a_{n-k+1} + c_ka_{n-k}\\ &= c_1a_{n-1} + c_2a_{n-2} + \cdots + c_{n-2}a_2 + c_{n-1}a_1 + c_na_0\end{aligned}$$

for all integers $n \geq k$. For each integer n such that $1 \leq n \leq k-1$, let

$$d_n = \sum_{j=1}^{n} c_j a_{n-j},$$

which is purely determined by the a_i and c_i. We conclude that

$$A(x)(1 - C(x)) = A(x) - A(x)C(x) = a_0 + (a_1 - d_1)x + (a_2 - d_2)x^2 + \cdots + (a_{k-1} - d_{k-1})x^{k-1}$$

is the desired polynomial $B(x)$. Indeed, it is of degree at most $k-1$, and its coefficients are determined by the a_i and c_i.

Solution 12.12. Using the formula from Theorem 12.11, and the fact that all elements of $[p]$, except p, are coprime to p, the answer is

$$\frac{1}{p} \cdot \sum_{i=1}^{p} k^{\gcd(p,i)} = \frac{1}{p} \cdot \left(\underbrace{k+k+\cdots+k}_{p-1 \text{ copies of } k} + k^p \right) = \frac{(p-1)k+k^p}{p}.$$

As a consequence, this means $(p-1)k+k^p$ is divisible by p, so

$$(p-1)k+k^p \equiv 0 \pmod{p} \implies k^p \equiv k \pmod{p}$$

for all positive integers k. The positive integers cover all residues classes, so all non-positive integers get covered as well. This is Fermat's little theorem.

Solution 12.13. The possible permutations of three vertices $\{1,2,3\}$ are:

$$e, (1\ 2)(3), (1\ 3)(2), (1)(2\ 3), (1\ 2\ 3), (1\ 3\ 2).$$

In order to use Burnside's lemma, let us count how many graphs each permutation fixes:

1. The identity permutation e fixes all 2^6 directed graphs because there are 6 possible directed edges, each of which are independent of each other.

2. Each 2-cycle or transposition fixes 2^3 directed graphs: for example, $(1\ 2)(3)$ fixes a graph if and only if both or neither of the $1 \to 2, 2 \to 1$ edges exist, both or neither of the $2 \to 3, 1 \to 3$ edges exist, and both or neither of the $3 \to 2, 3 \to 1$ edges exist.

3. Each 3-cycle fixes 2^2 directed graphs: for example, $(1\ 2\ 3)$ fixes a graph if and only if all or none of the edges $1 \to 2, 2 \to 3, 3 \to 1$ exist, and all or none of the edges $3 \to 2, 2 \to 1, 1 \to 3$ exist.

Therefore, the answer is

$$\frac{2^6 + 3 \cdot 2^3 + 2 \cdot 2^2}{6} = 16.$$

Solution 12.15. The right side will be shown to be the version of the left side with like terms collected. Each term $k^{\gcd(n,i)}$ is of the form k^d for some positive divisor $d \mid n$. By Lemma 12.14, for each $d \mid n$, there are $\varphi\left(\frac{n}{d}\right)$ elements i of $[n]$ such that $\gcd(n,i) = d$. Therefore,

$$\sum_{i=1}^{n} k^{\gcd(n,i)} = \sum_{d \mid n} \varphi\left(\frac{n}{d}\right) k^d = \sum_{ab=n} \varphi(a) k^b = \sum_{d \mid n} \varphi(d) k^{\frac{n}{d}}.$$

List of Symbols

Arithmetic

$\mathbb{Z}$	integers
$\mathbb{Z}_+$	positive integers
$\mathbb{Z}_{\geq 0}$	non-negative integers
$\mathbb{Q}$	rational numbers
$\mathbb{Q}_+$	positive rationals
$\mathbb{Q}_{\geq 0}$	non-negative rationals
$\mathbb{R}$	real numbers
$\mathbb{R}_+$	positive reals
$\mathbb{R}_{\geq 0}$	non-negative reals
$\mathbb{C}$	complex numbers
$\pm$	plus or minus
$<, >$	strict inequality
$\leq, \geq$	non-strict inequality

Constants

$\zeta_k = e^{\frac{2k\pi}{m}i}$	m^{th} root of unity
e	Euler's constant
ϕ	the golden ratio
B_n	the nth Bell number
$p(n)$	number of partitions of n
F_n	the nth Fibonacci number
C_n	the nth Catalan number
L_n	the nth Lucas number
P_n	the nth Pell number

Functions

$\lfloor \cdot \rfloor$	floor function
$\lceil \cdot \rceil$	ceiling function
max	maximum function
min	minimum function
det	determinant
Id_S	identity function on S
$f \circ g$	function composition
$f * g$	Dirichlet convolution
μ	Möbius function
$n!$	factorial
$P(n, k)$	number of k-permutations of n
$\binom{n}{k}$	binomial coefficient
$\binom{n}{n_1, n_2, \ldots, n_k}$	multinomial coefficient
$\left\{ n \atop k \right\}$	Stirling partition number
$L(n, k)$	Lah number
J_s	s^{th} Jordan totient function
φ	Euler's totient function
sin	sine
cos	cosine

Graphs

$\mathcal{V}$	vertex set of a graph
$\mathcal{E}$	edge set of a graph
$\mathcal{F}$	face set of a planar embedding
$\deg v$	degree of vertex v
$\deg f$	degree of face f
χ	Euler characteristic of a graph

Miscellaneous

$\exists$	existential quantifier
$\forall$	universal quantifier
$(a_i)_{i \in I}$	sequence indexed by I
$\sum$	summation notation
$\prod$	product notation
$a \sim b$	equivalence relation
$a \approx b$	equipotence

Probability

$\mathbb{P}$ probability distribution

$\mathbb{P}(A \mid B)$ conditional probability

$\mathbb{E}(f)$ expected value

Ω sample space

ω atomic events

$\mathcal{P}(\Omega)$ event space

Sets

$\varnothing$ empty set

$\in$ element of

$\notin$ not element of

$[n]$ $\{1, 2, \ldots, n\}$ for positive integers n

$[n]^*$ $\{0, 1, 2, \ldots, n\}$ for non-negative integers n

S^c or $\overline{S}$ set complement

$\cup$ set union

$\cap$ set intersection

$A \backslash B$ set difference

$A \times B$ Cartesian product of sets

A^n $\underbrace{A \times A \times \cdots \times A}_{n \text{ copies of } A}$

$\{0,1\}^n$ set of binary n-tuples

$\langle \ldots \rangle$ multiset notation

$\mathcal{P}(A)$ power set

$\subseteq$ subset

$\subsetneq$ proper subset

$\supseteq$ superset

$\mathrm{Orb}(x)$ orbit

X/G set of orbits

$\mathrm{Stab}(x)$ stabilizer

$\mathrm{Fix}(g)$ fix

Bibliography

"When we can't think for ourselves, we can always quote."

– *Ludwig Wittgenstein*

"Have you ever observed that we pay much more attention to a wise passage when it is quoted than when we read it in the original author?"

– *Philip Gilbert Hamerton*

"Everything of importance has been said before by somebody who did not discover it."

– *Alfred North Whitehead*

[1] Dušan Djukić et al. *The IMO Compendium, second edition.* Springer, 2011, pp. 6–7.

[2] Neil Calkin and Herbert S. Wilf. "Recounting the Rationals". In: *The American Mathematical Monthly* 107:4 (2000), pp. 360–363.

[3] H. S. M. Coxeter. *Introduction to Geometry, second edition.* John Wiley and Sons, 1969, pp. 152–154.

[4] Arthur Engel. *Problem-Solving Strategies.* Springer, 1998, pp. 96–97.

[5] Paul Hoffman. *The Man Who Loved Only Numbers.* Hachette Books, 1998, pp. 133–140.

[6] James R. Munkres. *Topology, second edition.* Prentice Hall, 2000, pp. 39–43.

[7] N. J. A. Sloane. *Catalan numbers.* URL: `https://oeis.org/A000108`.

[8] Richard P. Stanley. *Catalan Numbers.* Cambridge University Press, 2015.

[9] Richard P. Stanley. *Enumerative Combinatorics: Volume 1, second edition.* Cambridge University Press, 2011, pp. 71–79.

Index

"We raise to degrees (of wisdom) whom We please: but over all endued with knowledge is one, the All-Knowing."

– *Qur'an 12:76*

abelian group, 133
addition principle, 13
almost collectively exhaustive, 104
almost never, 104
almost surely, 104
almost-partition, 104
André's reflection principle, 103
average, 12

balls and boxes, 68
Bayes's rule, 110
Bell number, 75
Bertrand's ballot problem, 102
Bertrand's paradox in probability, 105
bijection principle, 8
binary tree, 124
Binet's formula for Fibonacci numbers, 119
binomial coefficient, 34
binomial distribution, 112
binomial theorem, 47
block-walking, 39
Bonferroni's inequalities, 63
Boole's inequality, 104
Burnside's lemma, 136

Calkin-Wilf tree, 8
Cantor snake, 8
cardinality of a set, 7
casework, 14
Cassini's identity, 119
Catalan numbers, 121
    formula, 123
chain rule for events, 107
characteristic polynomial, 128
chromatic polynomial, 97
clique, 89
closed walk, 80
combination, 33
combinatorial identity, 38
complement, 15
complementary counting, 15
    symmetric, 16
complementary probability, 100
complete graph, 89
composition, 70
    weak, 70
compositions, 73
conditional probability, 106
continuous probability, 105
correspondence principle, 31
countable, 8
countably infinite, 8
counting, 7
Coxeter, H. S. M., 87
cycle decomposition notation, 137

deletion-contraction recurrence, 97
dependence number, 29
derangement, 65
determinant, 142
    multiplicative, 143
discrete intermediate value theorem, 102
distinct, 68
distinguishable, 68
division principle, 31
double counting, 38
Dyck word, 121

edge colouring, 89
equipotent, 5
Erdős-Szekeres theorem, 92
Euler's constant, 66
Euler's partitions theorem, 53
Euler's totient function, 65
events, 98
expected value, 112
    linearity, 113
    multiplicative, 115

Fáry's theorem, 83
factorial, 30
Fermat's last theorem modulo p, 94
Fibonacci numbers, 118
finite Schröder-Bernstein theorem, 21
finite set, 6
fix, 134
fixed point, 66, 114
forming committees, 42

friends and strangers, 88

generating function, 49
  Fibonacci numbers, 120
geometric probability, 105
golden ratio, 119
graph, 78
  bipartite, 86
  complete, 79
  connected, 80
  cycle, 80
  directed, 78
  edges, 78
  finite, 78
  infinite, 78
  multigraph, 78
  path, 80
  regular, 86
  simple, 78
  subgraph, 79
  triangle-free, 85
  undirected, 78
  vertex degree, 79
  vertices, 78
  walk, 80
  with or without loops, 79
group, 133
group action, 134

handshaking lemma, 79
hockey stick identity, 44
  reverse, 45
hypergeometric distribution, 113

IMO Compendium, 132
independent events, 107
  mutually, 107
  pairwise, 107
indistinguishable, 68
infinite set, 6
initial segment, 121
injection-surjection lemma, 20

j-subset, 7
Jordan totient function, 65

k-set, 7
k-to-1 correspondence, 31
kappa function, 72
Kolmogorov axioms, 99
Kuratowski's theorem, 86

Lah numbers, 74
lattice point, 38
law of total probability, 108
leaf, 79
linear recurrence relation, 125
  (non-)homogeneous, 125
  depth, 125
  initial conditions, 126
list, 10, 91
Lucas numbers, 130

Markov's inequality, 114
matrix, 142
  identity, 142
  multiplication, 142
maximum-minimums identity, 63
median, 12
mode, 12
monochromatic, 89
monotonic sublist, 91
Monty hall problem, 110
multinomial coefficient, 36
multinomial set, 36
multinomial theorem, 49
multiplication principle
  dependent, 29
  independent, 26
multiset, 11
  multiplicity, 11
  pairwise disjoint, 13
  support, 11
multiset coefficient, 71
mutually exclusive, 104
  almost, 104

necklace, 137
necklaces, 137

OEIS, 121
one-to-one correspondence, 5
orbit, 134
orbit-stabilizer theorem, 135

Pólya enumeration theorem, 136
palindrome, 11
partition, 53
  function, 76
  generalized, 13
  integer, 76
  ordinary, 13
  probability, 104
  set, 75
Pascal's identity, 40
Pascal's triangle, 40, 118
Peano axioms, 125
Pell numbers, 130
permutation, 28, 35
  circular, 32

indecomposable, 46
k-permutation, 28
linear, 32
multinomial set, 37
permutation tree, 29
pigeonhole principle
optimality, 24
reverse, 21
strong, 23
strong reverse, 24
pigeonhole-principle, 21
planar graph, 83
bounded face, 83
embedding, 83
Euler's formula, 83
face degree, 84
planarity criterion, 84
unbounded face, 83
Platonic solid, 86
classification, 86
preimage, 35
preimage principle, 19
principle of inclusion-exclusion, 60
symmetric, 61
three sets, 18
two set, 17
probabilistic method, 115
probability, 98
monotonicity, 100
probability space, 98
distribution, 98
sample space, 98
projection map, 26
properties of determinants, 143

Ramsey number, 89
lower bound, 115
upper bound, 90, 93
Ramsey's theorem for c colours, 92
Ramsey's theorem for two colours, 89
random variable, 111
independent, 114
indicator variable, 112
range, 12
rationals are countable, 8
recursive relation, 39
unrestricted, 125
regular polyhedron, 86
Rule of Sarrus, 143

Schlegel diagram, 86
Schur's theorem, 94
skeleton, 86
stabilizer, 134
Stanley, Richard, 121, 123
Stirling partition number, 75
sublist, 91
subtraction principle, 15
surjections, number of, 67
symmetrically dependent, 29

Thomsen graph, 86
three utilities problem, 86
Thue's lemma, 23
topological ordering, 39
tree, 80
Cayley's formula, 83
number of, 82
properties, 80
triangulation, 123
triangulation of a polygon, 123
tuple, 10
twelvefold way, 68

uniform probability, 99
union bound, 18

Vandermonde matrix, 128
Vandermonde's identity, 45, 51

weak compositions, 70
well-ordering principle, 20

About the Author

"If you would be a real seeker after truth, it is necessary that at least once in your life you doubt, as far as possible, all things."

– *René Descartes, Principia Philosophiae*

"He who has a why to live for can bear almost any how."

– *Friedrich Nietzsche, Twilight of the Idols*

Samer Seraj is the owner of Existsforall Academy, where he enjoys teaching mathematics to his students. He earned his undergraduate degree in mathematics from the University of Toronto with membership at Trinity College. He also obtained a Masters degree in international affairs from King's College London. He is the author of several mathematical textbooks and has published research papers in journals. His vacation time is often spent on marking math contests and giving invited lectures, including formerly as a trainer of Math Team Canada for the International Mathematical Olympiad.

`https://existsforall.com/`

Made in United States
Cleveland, OH
06 November 2026

47669891R00101